W9-DJA-923

Chemistry Matters!

ATOMS AND MOLECULES

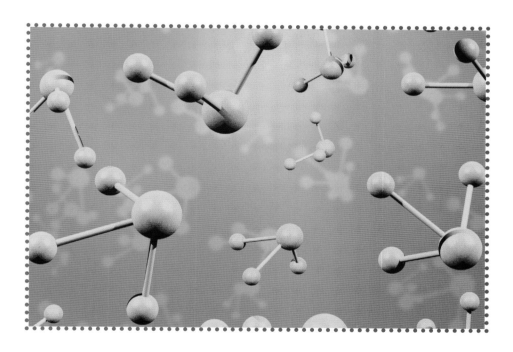

Volume 1

Tom Jackson

GROLIER

an imprint of

www.scholastic.com/librarypublishing

About this set

Chemistry Matters! provides an intelligent and stimulating introduction to all areas of modern chemistry as reflected in current middle school and high school curricula. This highly visual set clearly explains principles and applications using dramatic photography and annotated artwork. Carefully chosen examples make the topic fun and relevant to everyday life. Panels detail key terms, people, events, discoveries, and technologies, and include "Try This" features, in which readers are encouraged to discover principles for themselves in safe step-by-step experiments at home or school. "Chemistry in Action" boxes give everyday examples of chemical applications.

First published in 2007 by Grolier, an imprint of Scholastic Library Publishing
Old Sherman Turnpike
Danbury, Connecticut 06816

© 2007 The Brown Reference Group plc

Volume ISBN 0-7172-6195-6; 978-0-7172-6195-6
Set ISBN 0-7172-6194-8; 978-0-7172-6194-9

Library of Congress Cataloging-in-Publication Data
Chemistry matters!
    v. cm.
  Includes bibliographical references and index.
  Contents: v.1. Atoms and molecules—v.2. States of matter—v.3. Chemical reactions—v.4. Energy and reactions—v.5. The periodic table—v.6. Metals and metalloids—v.7. Nonmetals—v.8. Organic chemistry—v.9. Biochemistry—v.10. Chemistry in action.
  ISBN 0-7172-6194-8 (set : alk. paper)—ISBN 0-7172-6195-6 (v.1 : alk. paper)—ISBN 0-7172-6196-4 (v.2 : alk. paper)—ISBN 0-7172-6197-2 (v.3 : alk. paper)—ISBN 0-7172-6198-0 (v.4 : alk. paper)—ISBN 0-7172-6199-9 (v.5 : alk. paper)—ISBN 0-7172-6200-6 (v.6 : alk. paper)—ISBN 0-7172-6201-4 (v.7 : alk. paper)—ISBN 0-7172-6202-2 (v.8 : alk. paper)—ISBN 0-7172-6203-0 (v.9 : alk. paper)—ISBN 0-7172-6204-9 (v.10 : alk. paper)
 1. Chemistry—Encyclopedias.
  QD4.C485 2007
  540—dc22
                          2006026209

For The Brown Reference Group plc
Project Editor: Wendy Horobin
Editors: Paul Thompson, Tom Jackson,
            Susan Watt, Tim Harris
Designer: Graham Curd
Picture Researchers: Laila Torsun, Helen Simm
Illustrators: Darren Awuah, Mark Walker
Indexer: Ann Barrett
Design Manager: Sarah Williams
Managing Editor: Bridget Giles
Production Director: Alastair Gourlay
Editorial Director: Lindsey Lowe
Children's Publisher: Anne O'Daly

Academic Consultants:
Dr. Donald Franceschetti, Dept. of Physics,
  University of Memphis
Dr. Richard Petersen, Dept. of Chemistry,
  University of Memphis

Printed and bound in Singapore.

Contents

1 What Is Matter?

Everything in the universe is made from matter. Matter is made up of tiny building blocks called atoms. Chemistry is the science that investigates how atoms are organized to make the huge variety of substances we see around us.

Everything around you is made of matter. The pages of this book, the air you breathe, and even your body are made from the same building blocks. These building blocks do not make up just the things on Earth. Everything in the universe—the Sun, the billions of other stars, rocks, and clouds of dust, are made of them, too.

INTRODUCING ATOMS

The building blocks of matter are called atoms. Atoms are tiny and are far too small to see. About 125 million atoms lined up in a single row would be an inch long. However, not all atoms are the same. There are about 90 different types in nature. Atoms come in different sizes and masses and have many properties.

On Earth there are three main types of matter: solids, liquids, and gases. The Sun is a fourth type of matter called a plasma. Plasmas behave in a similar manner to gases but are very much hotter. The Sun's plasma is made up of the simplest of all atoms—hydrogen.

Atoms group together to make the objects and other materials around us. Matter that contains just one type of atom is called an element. For example, a gold nugget contains only gold atoms. Other elements include carbon, iron, aluminum, sulfur, and oxygen.

MOLECULES

Elements are unique materials because they cannot be broken up into simpler ingredients. However, not everything in the universe is made of pure elements. Most things contain combinations of the atoms of several different elements.

A Closer LOOK

Density

One of the main properties of matter is density. Density is a measure of how much matter a substance has packed inside it. Density is measured by comparing an object's mass with its volume. Certain materials contain a lot of large, heavy atoms or have molecules arranged very closely together. This packing makes even small volumes of such materials very heavy. Scientists describe these substances as having a high density. For example, lead is a very dense metal. It is used to make weights. Other substances have small atoms that are spread out. Helium gas is one of the least dense substances known. It is used inside airships to make them light enough to float in the air.

▲ Airships are filled with helium to make them lighter than air. In the past, hydrogen was used to fill airships, but hydrogen is very flammable. Its use led to a number of airships being destroyed by fire, killing the passengers. Helium is heavier than hydrogen but is not flammable and is therefore much safer.

Combinations of different types of atoms are called compounds.

When the atoms group together, they make a structure called a molecule. A molecule has a unique shape and size. This gives a material certain properties, such as making it hard or bendable. Compounds frequently have properties that are different from those of the elements included in the molecule. For example, as an element, sodium is a soft metal that reacts violently with water. Chlorine is a highly reactive gas. When combined as common table salt, sodium and chlorine form a glassy crystal that is stable, safe, and unreactive at room temperature.

Key Terms

- **Atom:** The smallest independent building block of matter.
- **Element:** A material that cannot be broken up into simpler ingredients.
- **Molecule:** Two or more joined atoms that have a unique shape and size.

▼ *Erupting volcanoes show matter in all three states—the solid rock that forms the volcano, the liquid magma that pours out of it, and the gases that are blasted into the atmosphere.*

STATES OF MATTER

All matter has three basic forms, or states: solid, liquid, and gas. There is a fourth state of matter, called plasma, but most of the material on Earth tends to exist as a solid, a liquid, or a gas. A solid, such as stone, is hard and has a fixed

TRY THIS

Ice, water, steam

Volume is the measurement of the amount of space a substance occupies. So does volume change with the state of matter? Take an empty ice cube tray and fill the compartments with water. Fill each compartment with as much water as possible without spilling any.

Freeze the water in your freezer. Now leave the tray in a warm spot indoors. After a few minutes, the ice will have melted into water. Has any water spilled into other compartments? If you have been careful, no water should have spilled. Liquids have more or less the same volume as their solid states. Water is an unusual liquid because it expands slightly when it freezes. However, when the ice melts the water should still all fit in the single compartment.

Now place the tray outside in a place in the sun. After a few hours, most of the water will have evaporated into water vapor—it has become a gas. Gases do expand when they form and the molecules move farther apart. This experiment shows that molecules are more or less as tightly packed in a solid as they are in a liquid. However, in a gas they are free to spread out completely.

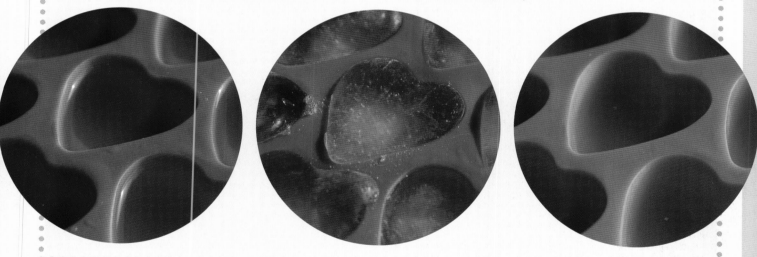

shape. A liquid, such as water, takes on the shape of its container. It can flow from one shape into another, such as when it is poured into a new holder or pumped through a pipe. In this respect, a gas is similar to a liquid because it too takes on the shape of its container. However, liquids always sit at the bottom of a container. Gases are different. They fill all the available space by spreading out so their atoms or molecules are evenly dispersed throughout the entire container. In normal conditions, such as those inside a house, each material will be in a certain state. Some are solids, such as metals and plastics. Some are liquids, such as water or fruit juice. Others are gases, such as the air. The state of each material depends on how its atoms and molecules are arranged.

▲ *These three pictures show how when water in an ice tray (left) is frozen it expands (center). When the ice is left to melt in the sun (right), some of the water evaporates and escapes as a gas.*

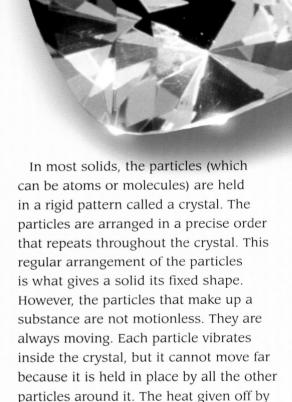

In most solids, the particles (which can be atoms or molecules) are held in a rigid pattern called a crystal. The particles are arranged in a precise order that repeats throughout the crystal. This regular arrangement of the particles is what gives a solid its fixed shape. However, the particles that make up a substance are not motionless. They are always moving. Each particle vibrates inside the crystal, but it cannot move far because it is held in place by all the other particles around it. The heat given off by a solid is a measure of how much its atoms are moving.

In some solids the particles are not held in rigid patterns but are arranged in an irregular way. These solids are called amorphous solids—they have no regular structure. In amorphous solids, strong bonds between the particles hold them in position. Like crystalline solids, the particles in amorphous solids vibrate constantly. Plastics and wax are common examples of amorphous solids.

▲ A diamond is a crystalline form of carbon. Its atoms are held in a rigid pattern. It is this structure that allows a diamond to be cut and polished into a sparkling gemstone.

▶ In some places, such as Yellowstone Park, water comes out of the ground boiling hot. Like other liquids, water boils at a certain temperature. In the case of water that temperature is 212°F (100°C). The water molecules are vibrating so strongly that some are turning into vapor and escaping from the surface as steam.

A Closer LOOK

States of matter

▼ **Solid:** The particles are held strongly together. This example below is a rigid crystal structure. The particles can move but only vibrate back and forth inside the crystal.

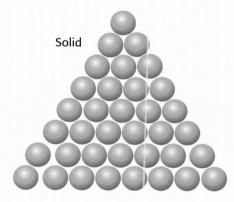

Solid

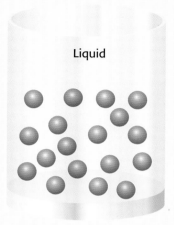

Liquid

▲ **Liquid:** The particles are less tightly packed than in a solid. They can move past each other, so a liquid can flow into different shapes.

▼ **Gas:** The particles are not joined to any other. All move independently of each other. A gas will spread out to fill all available space.

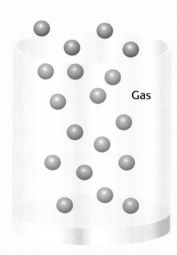

Gas

MELTING AND BOILING

A material can change from one state to another by being heated or cooled. Solids melt into liquids, and liquids evaporate into gases. In the opposite direction, gases condense into liquids, which then freeze into solids.

When a solid is heated, the particles begin to vibrate more strongly and move farther out of position. At a certain temperature, the particles inside the solid begin to break away from each other. The solid's shape breaks down, and the particles can now move around more freely. The substance has become a liquid. The temperature at which a substance turns from solid to liquid is called its melting point.

The particles in a liquid are still connected to some of the others. If a liquid is heated further, the particles

Key Terms

• **Gas:** State in which particles are not joined and are free to move in any direction.
• **Liquid:** State in which particles are loosely bonded and are able to move freely around each other.
• **Solid:** Matter in which particles are held in a rigid arrangement.
• **State:** The form that matter takes—either a solid, a liquid, or a gas.

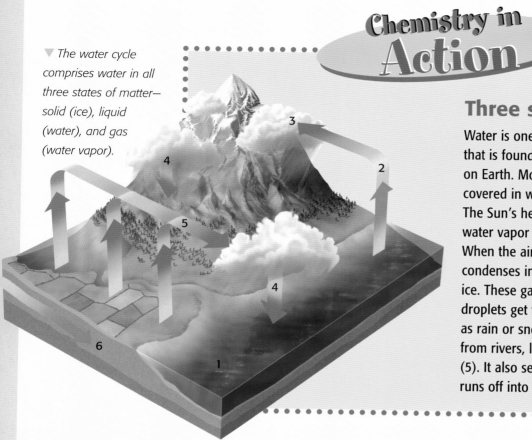

▼ The water cycle comprises water in all three states of matter—solid (ice), liquid (water), and gas (water vapor).

Chemistry in Action

Three states of water

Water is one of the very few substances that is found naturally in all three states on Earth. Most of Earth's surface is covered in water, forming the oceans (1). The Sun's heat evaporates seawater into water vapor that mixes with the air (2). When the air gets cold, the water vapor condenses into tiny droplets of liquid or ice. These gather as clouds (3). When the droplets get too large, they fall to Earth as rain or snow (4). Water also evaporates from rivers, lakes, and land to form clouds (5). It also seeps into the ground (6), or runs off into the sea.

begin to vibrate even more quickly. Eventually they are vibrating so much that all the bonds between the particles break, and a liquid turns into a gas. The temperature at which this change occurs is called the boiling point.

As a gas, the particles are free to move in any direction. They keep moving in a straight line until they hit other gas particles, or bounce off the solid surface of a container. With the particles bouncing in all directions, the gas soon spreads out to fill the container.

PHYSICAL PROPERTIES

When a substance is cooled, the processes work in the opposite direction. As gas particles cool down they move more slowly. When the particles hit each other,

◀ *A melting solid stays at a constant temperature until all the solid has melted. Here the ice and water remain at 32°F (0°C) until all the ice has melted.*

they are moving slowly enough to stay stuck to each other. Shapeless droplets of liquid are formed. As the liquid is cooled, its particles vibrate less strongly. They begin to connect together into the ordered structure of the solid crystal.

Each substance melts and boils at set temperatures, known as melting and boiling points. Melting and boiling points are physical properties. Physical

▲ *Distillation is a technique used to separate a mixture of liquids by the differences in their boiling points. As each liquid in the mixture reaches its boiling point, it evaporates into a gas and is collected at the top of the reactor. It is then cooled back into a pure liquid.*

properties include color and hardness, as well as melting and boiling points. For example, some solids are hard, such as stone and iron, while some are soft, such as soap. However, many metals are also easy to bend, especially when hot, but stone will crack. A thin sheet of copper or aluminum, for example, can be bent easily, but a slab of stone will shatter if you try to bend it.

 Salt crystals are a compound of two elements, sodium and chlorine, that have undergone a chemical reaction to form a new substance, sodium chloride.

These physical changes do not have any effect on the structure of the particles inside the substance. Whether they are joined in a crystal or floating independently as a gas, the particles stay the same. However, substances have chemical properties, too. These properties affect the way that particles reorganize themselves to make completely new substances.

CHEMICAL REACTIONS

In the right conditions, two or more different substances will react with each other to make completely new materials. During the reaction, the atoms inside the two original substances are rearranged to form new molecules. A chemical change has occurred because the original substances no longer exist. Instead they have transformed into one or more new materials.

The new molecules formed by the chemical reaction have a different shape and size from the original ones. They also contain different combinations of atoms. This new combination of atoms may have very different properties from those of the starting materials. For example, solid and liquid materials might react to form a gas.

COMPOUNDS AND MIXTURES

When atoms of two or more elements react to form a molecule, the new substance that is formed is called a compound. The compound contains atoms of several elements. When they are combined by a chemical reaction,

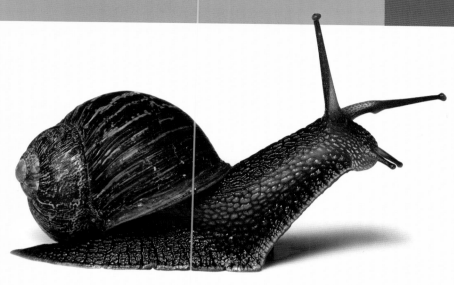

▲ *The shell of this snail is made of calcium carbonate. Calcium compounds are key ingredients for many animals in making bones, teeth, and shells.*

the compound formed generally has very different properties, both physical and chemical, from those elements.

For example, calcium carbonate ($CaCO_3$) is a compound that consists of calcium, carbon, and oxygen. Calcium is a soft metal, carbon is a nonmetal, and oxygen is a gas under ordinary conditions. Yet, when combined by a chemical reaction, they produce a crystalline substance that is used by some animals to form a hard shell. Calcium carbonate occurs in nature within the rocks known as limestone, chalk, and marble.

A Closer LOOK

Element, mixture, compound

Substances exist in different forms in nature. At the most basic level are substances known as elements. Elements consist of one type of atom. These atoms may be single atoms or molecules containing a number of atoms. Mixtures occur when molecules of different substances mingle together but do not combine physically or chemically. The ingredients of a mixture can be separated from each other. Compounds are the result of a chemical reaction between two or more substances and can only be separated into individual elements by chemical methods.

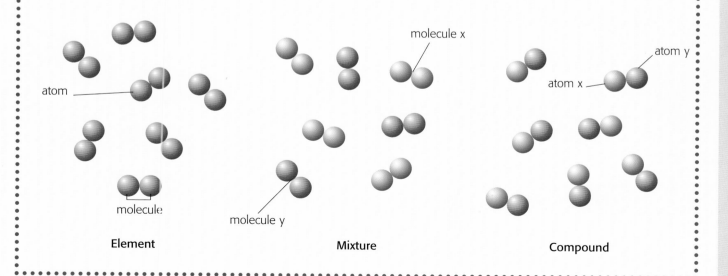

TRY THIS

Mixing mixtures

Mix a small amount of sand with a similar amount of salt. No matter how thoroughly you mix the two, you can always see the different grains of salt and sand. This is an example of a heterogeneous mixture. Now tip the mixture into a large glass of warm water and give it a stir. Let the grains settle. What can you see? The sand sinks to the bottom, but the salt grains have disappeared. They have dissolved into the water and have formed a homogeneous mixture—a solution.

▼ The sand and salt grains below are mixed together but each remains identifiable.
▷ However, when put into a glass of water (right) the sand sinks and the salt dissolves.

Compounds are different from mixtures. The calcium, carbon, and oxygen atoms are not randomly mixed together to make calcium carbonate crystals. Each calcium carbonate molecule is formed by a chemical reaction that joins one calcium, one carbon, and three oxygen atoms together. When joined, the atoms form a completely different substance.

TYPES OF MIXTURES

Inside a mixture of materials each substance still behaves as if it were in a pure state. For example, pebbles and water can be mixed together. However, they can also easily be separated from each other. When the mixture is heated, for example, the water boils away leaving the solid stones behind. Or the stones can simply be lifted out of the water. Mixing the two things has not changed either substance's physical properties. However, when a compound is formed, the substances are combined by a chemical reaction. The compound has a new set of properties and it cannot easily be separated into the original starting substances.

There are two main types of mixtures: heterogeneous and homogeneous. Heterogeneous mixtures contain two or more different materials. These materials are spread unevenly throughout the mixture. A bowl of breakfast cereal is a heterogeneous mixture because it is easy to see the different materials—the flakes and milk—that the bowl contains.

It is harder to determine what is included in a homogeneous mixture.

▲ This breakfast cereal is an example of a heterogeneous mixture. It is easy to see all the ingredients because none of them changes when put together in a bowl.

That is because all the ingredients are evenly mixed up so you cannot identify one from the other. For example, butter is a homogeneous mixture of water and fat. Many homogeneous mixtures involve water. Another example is a cup of coffee. The molecules in the coffee granules, which give the drink its taste and color, are evenly mixed up with the water molecules. When a spoonful of sugar is added, the sugar molecules also become part of the homogeneous mixture. It is impossible to tell the different ingredients apart.

Chemists describe a mixture like this as a solution. The water is called the solvent, and the coffee and sugar are called the solutes because they dissolve in the solvent.

Key Terms

- **Compound:** Matter made from more than one element and that has undergone a chemical reaction.
- **Heterogeneous mixture:** A mixture in which different substances are spread unevenly throughout.
- **Homogeneous mixture:** A mixture in which one substance has dissolved or been completely mixed into another.
- **Mixture:** Matter made from different types of substances that are not physically or chemically bonded together.

See Also ...

- *What Is a Chemical Reaction? Vol. 3: pp. 4–9.*
- *Energy in Chemical Reactions, Vol. 4: pp. 4–17.*

2 Introducing Elements

The elements are the basic substances in nature. They cannot be broken down into simpler substances. Each element has its own type of atom of a particular size and mass (amount of matter) and with certain chemical properties.

Not all atoms are the same. There are 92 types of atoms that occur naturally on Earth, and they make up the most basic substances called elements. Three-quarters of the elements are metals. Ten other elements are gases in normal conditions, while just two (including one of the metals) are liquids. Some elements will react with just about all the others, while a few rarely react at all.

It is possible that more than 92 elements occur elsewhere in the universe, but these only last for a short time before breaking down into more stable elements. None of these unstable elements exist naturally on Earth any more. Scientists have made some of them in laboratories, but they can only produce very tiny amounts at a time and they quickly break down.

Bismuth is a rare metallic element with a regular crystalline structure. Its pattern of interlocking squares and rainbow colors makes it one of the most distinctive elements.

Chemistry in Action

Earthy elemental facts

The most common element in the universe is hydrogen. Hydrogen atoms are the smallest and simplest of all atoms. Three-quarters of all matter in the universe is made of hydrogen.

The most common element on Earth is iron. Much of the planet's core is made of this metal. However, the most common elements on the surface of Earth are silicon and oxygen. For example, silicon dioxide is the main compound in sand and is also found in most rocks. Other elements are very rare indeed. For example, there is just 1 ounce (28 g) of astatine in all of Earth's rocks put together.

▶ *Earth is made of several layers comprising different elements and compounds. Iron is the most common element on Earth and may be either solid or liquid, depending on pressure and temperature.*

Solid mantle made of iron, magnesium, aluminum, silicon, and compounds of oxygen and silicon. Divided into upper mantle and lower mantle.

Liquid iron and nickel outer core.

Solid iron and nickel inner core.

Continental and oceanic floor crust made of aluminum, calcium, silicon, and oxygen compounds.

DISCOVERING ELEMENTS

People have long understood that certain basic substances can be combined to make new and completely different substances. Today, chemists understand how atoms are constructed and what makes each element different from the next. However, before science provided these explanations, people had very different ideas about elements.

For many hundreds of years, people thought that everything in the world was made from just four elements: earth, fire, water, and air. The way these elements were supposed to work had more to do with magic than science. Nevertheless the idea of elements as fundamental substances was not incorrect, it was just that nobody had discovered any of the true elements.

This book on magic was published in 1658. In the corners are representations of the "four elements": fire, air, earth, and water. Clockwise from top left, fire is represented by a salamander, air by birds and wind, water by sea creatures, and earth by animals.

THE BIRTH OF CHEMISTRY

The first people to investigate how materials could be changed into other substances were not chemists but alchemists. Alchemists are first recorded working in Egypt and China about 2,500 years ago. Alchemists were not scientists and much of what they did is often remembered as the work of wizards and witches. They made potions and remedies and thought that matter could be transformed using magic.

A Closer LOOK

Gibbering alchemists

Alchemists were very different from modern chemists. Chemists are scientists and they share their discoveries with others. Chemists check each other's discoveries to make sure they are correct and then use them to learn more about how atoms and molecules behave.

In contrast, the main goal of an alchemist was to find one of three things: the elixir, a drink that could make a person live forever; the panacea, a medicine that could cure all illnesses; and the philosopher's stone, which could turn any metal into gold. Obviously these discoveries would have made an alchemist hugely powerful. As a result alchemists preferred to keep their work private. They recorded things in code, using strange symbols.

One of the most influential alchemists was the Arab Jabir ibn Hayyan, also known as Geber. His writings are very confusing and they often contradict each other. The word *gibberish*, meaning "to talk nonsense," comes from this man's name.

This engraving depicts an alchemist surrounded by the apparatus, instruments, and books of magic that he would use to carry out experiments and create potions. In the 18th century, as scientific knowledge increased, alchemy began to be replaced by chemistry.

Unlike chemists, alchemists did not carry out proper scientific experiments. They also did not understand many of the basic ideas of chemistry, such as the difference between a compound and a mixture (*see* pp. 12–15). However, they did make some important discoveries. For example, alchemists began to understand that there were many more than just four elements. They identified several of the metal elements, such as mercury, iron, and gold. Alchemists also correctly thought that sulfur, arsenic, and other nonmetals were elements. They began to use symbols for each of the elements, and modern chemists do the same. However, nobody really understood that elements were made of atoms and formed compounds until chemists began to investigate elemental properties in a scientific way.

INVISIBLE ATOMS

Atoms have only been detected by scientists in the last 100 years. However, people have been talking about atoms for thousands of years. The first people to think atoms existed were ancient Greek philosophers. Unlike alchemists, philosophers did not perform any experiments to understand matter. They did not use science to prove their ideas, either. Instead they came up with theories that seemed to match the way they observed nature working.

The word *atom* comes from the Greek word *atomos*, which means "indivisible." The first person to suggest that matter is made of atoms was Leucippus of Miletus,

▲ Mercury is a metallic element and is one of only two elements that are liquid at room temperature—the other is bromine.

Key Terms

- **Alchemist:** Person who attempted to change one substance into another using a combination of primitive chemistry and magic.

- **Four elements:** The ancient theory that all matter consisted of only four elements (earth, air, fire, and water) and their combinations.

Tools and Techniques

Seeing atoms

There is a way of making a picture of the surface of an atom. This requires a machine called a scanning tunneling microscope. The microscope has a metal probe with an ultrafine point. At the tip of of this point there is a single atom. The probe works by picking up an electric current from the atoms it is detecting. It can be used only to investigate substances that conduct electricity, such as metals. The probe is lowered to just above the surface of an object. When the tip of the probe moves close to an atom, the current running through it rises. The atoms of certain elements produce larger bursts of electricity than others. A computer uses the rises and falls in the current to create a picture of where all the atoms are on the surface.

▲ The yellow-shaded area of this scanning tunneling microscope image shows gold atoms resting on a green-shaded layer of carbon atoms.

who lived about 2,500 years ago. He thought atoms were all the same and could not be squeezed, stretched, or broken. Leucippus thought atoms had to exist because things were constantly changing in nature. However, he understood that something new could not be made from nothing, so he suggested that all changes were just atoms being rearranged. The atoms themselves could not be changed, only the way they were organized. Leucippus and his followers did not understand anything about how atoms are constructed or why they behave the way they do. However, their atom theory was correct in many ways.

SCIENTIFIC APPROACH

When chemists began to study substances in a scientific way, they began to realize that matter was indeed made of atoms but that not all atoms were the same. John Dalton (1766–1844) was an English scientist who made one of the greatest discoveries in chemistry. At the beginning of the 19th century, he noticed that when two types of gases were mixed together, they did not behave as a single cloud that filled its container. Leucippus had said that all things were made up of identical atoms, so why were the atoms in each gas behaving differently? Dalton saw that both gases expanded independently of each other, so they

were both spread evenly throughout the container. This simple observation proved that not all atoms were the same, as Leucippus had thought. The two gases must contain different types of atoms that behave differently from each other.

WEIGHTS AND MEASURES

By the early 19th century, scientists had identified about 25 elements. These included metals such as gold, mercury, and copper, which had been known for centuries. There were also new additions, including oxygen, which had been discovered a few years before Dalton's experiment. Dalton suggested that each element had its own type of atom.

▲ The Guggenheim Museum in Bilbao, Spain, is covered with a thin layer of the metallic element titanium. Titanium is one of the 25 elements that had been discovered by the early 19th century.

▶ English scientist John Dalton was the first person to propose that atoms of different elements do not all behave the same way.

The main difference Dalton could find between the various elements was their mass (how much matter the substance contains) and density (how much mass something has at a particular volume). He figured out how heavy an element was by reacting it with a fixed mass of another element. He then weighed what was produced and could calculate how heavy the first element was compared with the second. The extra mass of the new compound indicated how heavy the added atoms were. With this measurement, Dalton could begin to figure out how heavy each type of atom was in relation to all the others.

Dalton also showed that compounds formed from even proportions of elements. For example, a compound could contain equal amounts of two elements, or have twice as much of one than the other. However, the proportions of each ingredient were always whole numbers. You never find one atom of an element joined to one and a half atoms of another element.

SYMBOLS AND FORMULAS

Although he did not know it at the time, Dalton was describing how atoms are arranged inside a compound's molecules.

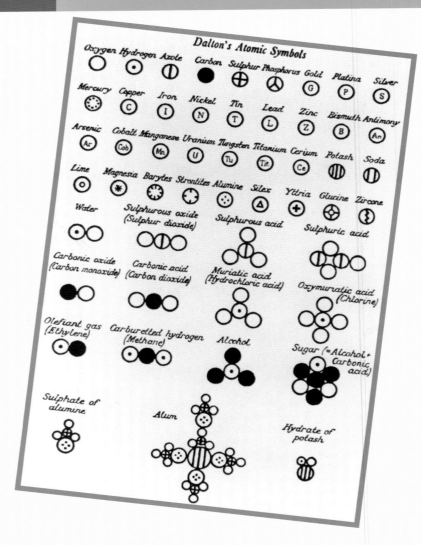

This table of elements, compounds, and chemical symbols was created by English scientist John Dalton in 1808. Each circular symbol represents one atom, and molecules are represented by combinations of these symbols. This system is no longer used but many of Dalton's ideas were essentially correct.

Key Terms

- **Chemical symbol:** The letters that represent a chemical, such as "Cl" for chlorine or "Na" for sodium.
- **Chemical formula:** The letters and numbers that represent a chemical compound, such as "H_2O" for water.

A simple compound such as sodium chloride (common salt) is produced when equal amounts of sodium and chlorine atoms react together. The proportions of each element, or ratio, is 1:1. However, other compounds have more complicated ratios of elements. For example, water is formed when hydrogen and oxygen react. Twice as many hydrogen atoms are needed than oxygen atoms. So the compound has a ratio of 2:1.

Chemists use these proportions to explain the exact ingredients of molecules. Like alchemists before them, chemists use symbols for each element. However,

A Closer LOOK

Molecules and formulas

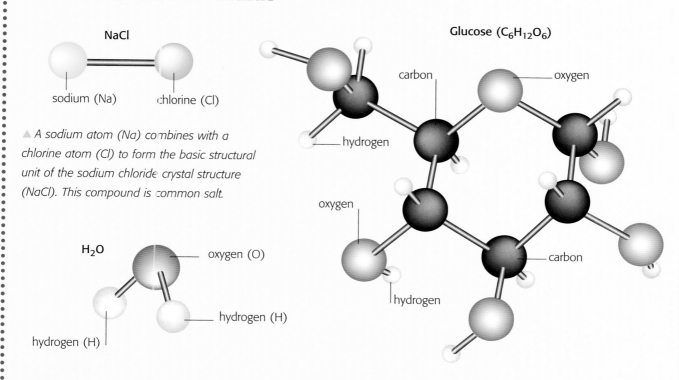

NaCl

sodium (Na)   chlorine (Cl)

▲ A sodium atom (Na) combines with a chlorine atom (Cl) to form the basic structural unit of the sodium chloride crystal structure (NaCl). This compound is common salt.

H_2O

oxygen (O)

hydrogen (H)

hydrogen (H)

▲ A single oxygen atom (O) combines with two hydrogen atoms (2H) to produce a water molecule (H_2O).

Glucose ($C_6H_{12}O_6$)

carbon

oxygen

hydrogen

oxygen

carbon

hydrogen

▲ Glucose is a type of sugar. This complex molecule is made of six carbon atoms (6C), twelve hydrogen atoms (12H), and six oxygen atoms (6O). Together these form $C_6H_{12}O_6$.

they have chosen ones that are much easier to understand. The symbol of hydrogen is H; chlorine is Cl; and oxygen has the symbol O. Some elements have less obvious symbols because they are taken from languages other than English. For example, sodium has the symbol Na, which comes from the Latin word *natrium*. Iron has the symbol Fe from *ferrum*, the Latin word for iron.

Chemists combine the symbols and the proportions to make chemical formulas.

These are a way of describing what proportions of elements a compound contains. For example, the formula for sodium chloride is NaCl. Water has the formula H_2O. The subscript 2 indicates that in each molecule of water, two hydrogen atoms are joined with one oxygen atom. More complicated compounds, such as glucose (a type of sugar), have large molecules containing many atoms. The formula for glucose is $C_6H_{12}O_6$.

See Also ...
• *Properties of Metals,* Vol. 6: pp. 4–13.
• *Atoms and Elements,* Vol. 5: pp. 4–9.

3 Looking at Atoms

An element's properties come from the way it is made. Each element is made up of particular atoms. Tiny atoms are themselves made up of even smaller particles. Each element has a unique number of these particles inside its atoms.

For hundreds of years, scientists thought that atoms were the smallest particles of matter. In the 20th century scientists began to realize that atoms were made of even smaller particles called subatomic particles. There are three types of subatomic particles in atoms. These are protons, neutrons, and electrons. All atoms are constructed using these particles, which are always arranged following the same rules. An element's physical and chemical properties are set

These curves and spirals show the paths of subatomic particles recorded by a machine called a particle detector.

by the number of particles in its atoms. For example, an element with many particles in its atoms will be very dense. However, an element with just a few particles in its atoms will have a low density. Similarly, the number of particles in an element's atoms governs how reactive the element is.

PROTONS

The inside of an atom is mostly empty space, and the particles inside it are incredibly small. At the center is a tiny core called the nucleus (plural, nuclei).

Chemistry in Action

Tiny particles

Subatomic particles are extremely small. One gram of electrons contains more than a 1,000 times more individual particles in it than there are stars in the entire universe. (Astronomers think there are about ten thousand billion billion stars in all.)

The largest atoms are about five millionths of a millimeter across. However, the nucleus, where most of the atom's mass is located, is just a few trillionths of a millimeter across. That is like a ping-pong ball positioned in the center of a sports stadium.

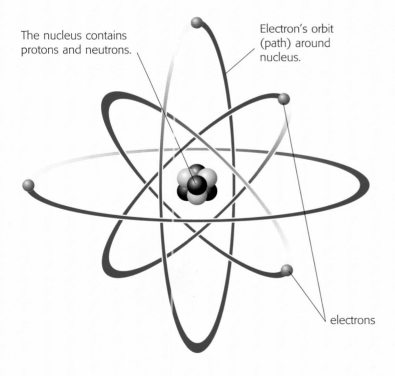

The nucleus contains protons and neutrons.

Electron's orbit (path) around nucleus.

electrons

▲ *An atom is mostly empty space. The protons and neutrons are gathered in the central region called the nucleus. The electrons orbit the nucleus.*

The nucleus contains the first type of particle: the proton. The name *proton* was first used to describe this particle around 1920 and comes from the Greek word for "first." The proton was given this name because it was the first particle to be found inside an atom. (Electrons were discovered in 1897 but they were not known to be part of atoms until after the discovery of protons.)

The number of protons in an atom defines the type of element. The simplest and smallest type of atoms—those of hydrogen—have just a single proton in their nucleus. Larger atoms have more protons in the nucleus. For example, the largest naturally occurring element is the metal uranium. Uranium atoms have 92 protons. The number of protons in an atom is called the atomic number. Each element has a unique number of protons

▶ This table shows the subatomic particles of the first 14 elements in the periodic table. The number of neutrons is not always the same as the number of protons. The atomic mass number is the sum of the atomic number and the number of neutrons.

	Number of protons (atomic number)	Number of neutrons	Number of electrons	Atomic mass number
Hydrogen	1	0	1	1
Helium	2	2	2	4
Lithium	3	4	3	7
Beryllium	4	5	4	9
Boron	5	6	5	11
Carbon	6	6	6	12
Nitrogen	7	7	7	14
Oxygen	8	8	8	16
Fluorine	9	9	9	18
Neon	10	10	10	20
Sodium	11	12	11	23
Magnesium	12	12	12	24
Aluminum	13	14	13	27
Silicon	14	14	14	28

▼ Aluminum has many common uses, such as in soft drinks cans. Each of the tiny aluminum atoms is made of a nucleus of 13 protons and 14 neutrons surrounded by 13 electrons.

in its nucleus. If two atoms have different atomic numbers, then they belong to different elements.

Protons have a positive electrical charge. This charge is a fundamental property of protons, and each one has an identical charge. Chemists describe the charge of each proton as +1.

A proton's charge is linked to the way it pushes and pulls on other particles inside the atom. Because of the protons in it, a nucleus always has a positive charge. The nucleus of a larger atom contains a lot of protons. The charges of these particles add up to make a positive charge that is stronger than that of a smaller nucleus with fewer protons.

NEUTRONS

With the exception of hydrogen, all elements have a second type of particle in their nuclei. These are called neutrons. Neutrons are slightly heavier than protons. However, they have no electrical charge—they are neutral. Because they have no charge, neutrons do not play much of a role in chemical reactions.

The simplest element to have neutrons in its atoms is helium. This has two protons in the nucleus and two neutrons.

Atomic structure

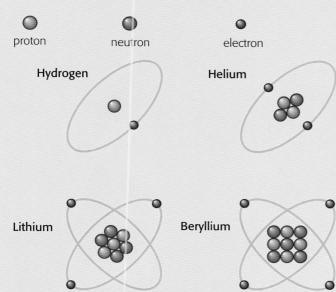

proton   neutron   electron

Hydrogen   Helium

Lithium   Beryllium

▲ The number of protons in the nucleus of an atom determines the type of element the atom is. The number of neutrons is usually similar to the number of protons but is not always the same. Protons and electrons have an electric charge. The protons are positively charged and the electrons are negatively charged. Neutrons are electrically neutral; they have no charge.

Key Terms

• **Atomic mass number:** The number of protons and neutrons in an atom's nucleus.
• **Atomic number:** The number of protons in a nucleus.
• **Electron:** A small, negatively charged subatomic particle that circles the nucleus.
• **Neutron:** A subatomic particle with no charge located in an atom's nucleus.
• **Nucleus:** The central part of an atom, made of protons and neutrons.
• **Proton:** A positively charged particle found in an atom's nucleus.
• **Subatomic particles:** Particles that are smaller than an atom.

The number of neutrons in larger atoms is also roughly the same as the number of protons. However, this varies a lot from element to element. The number of particles in an atom's nucleus—protons plus neutrons—is called the atomic mass number. For example, most hydrogen atoms have an atomic mass number of 1 (1 proton plus 0 neutrons) and most carbon atoms have atomic mass numbers of 12 (6 protons and 6 neutrons). The atomic mass number tells scientists how much matter is contained inside an atom and how heavy it is.

ELECTRONS

The third type of subatomic particle, the electron, is not located in the nucleus. Instead, electrons move around (orbit) the nucleus. Electrons are about 1,830 times lighter than a proton or neutron.

Despite their small size, electrons have a negative electrical charge of –1. This charge is exactly equal and opposite to the charge of each proton. Particles with opposite charges attract each other, and the negatively charged electrons are pulled toward the positively charged nucleus. This force keeps the electrons in position and holds the atom together.

The number of electrons in an atom is always the same as the number of protons. For example, hydrogen atoms have one electron, while helium atoms have two. Therefore, the positive charge of the protons is balanced exactly by the charge of the electrons. As a result whole atoms never have an overall charge.

VARYING NUMBERS

All the atoms of one element must have the same atomic number. However, atoms of the same element can have slightly different atomic mass numbers. That is because an element can have

▼ *The reddish color in the rocks of the Grand Canyon is because of iron in the rocks. Iron is a common constituent of many rocks. Iron atoms do not all have the same atomic mass number. Around 90 percent of iron atoms have a mass number of 56. The remaining 10 percent have different atomic mass numbers because they have different numbers of neutrons in their nucleus.*

Tools and Techniques

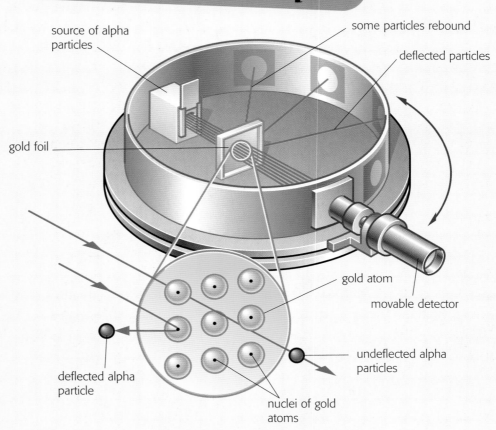

▶ *In Rutherford's experiment, a beam of alpha particles (helium nuclei) is fired at a thin film of gold foil. This arrangement is contained within a vacuum so that the alpha particles do not rebound from air particles. A movable detector swivels around the gold foil and measures the quantity of alpha particles that pass through the foil or that are deflected by the foil. Most pass straight through, but some are deflected and a small number rebound.*

source of alpha particles

some particles rebound

deflected particles

gold foil

gold atom

movable detector

deflected alpha particle

undeflected alpha particles

nuclei of gold atoms

Discovering atomic structure

Electrons were discovered in 1897, and protons were first isolated in 1910, though they were not given the name proton until around 10 years later. Neutrons were discovered in 1932. At first scientists thought that atoms were more or less solid arrangements of protons with the smaller electrons positioned in between to balance out the particles' different charges. However, in 1911 Ernest Rutherford (1871–1937), a New Zealander working in England, showed that protons were actually located in a tiny nucleus at the center of the atom. He discovered this by firing a beam of alpha particles at a very thin sheet of

gold. (An alpha particle is the nucleus of a helium atom that has lost its electrons. It has two protons and two neutrons and is positively charged.) Most of the alpha particles went straight through the gold, but some were deflected and others bounced back. Rutherford realized that the particles that were changing direction were being repelled by other positively charged particles inside the atom. Rutherford discovered that the beam of particles was only affected by very tiny areas of the gold sheet, areas that were smaller than the size of individual atoms. He realized that this meant that protons—the positively charged parts of atoms—only occupied a tiny area at the center of the atom, which he called the nucleus.

Chemistry in Action

Carbon dating

Some isotopes are radioactive—they decay into other isotopes or elements over time (*see* pp. 56–65). Radioactive isotopes can be used to tell how old an ancient object is. Different isotopes are used for different ages. For example, potassium-40 (K-40) is used to date rocks that are billions of years old. Carbon-14 (C-14) is used to calculate the age of once-living things, such as wood, bones, and even clothing made from natural material. Carbon is an essential ingredient of all living things. Scientists know that when living material is formed, it always contains a certain proportion of C-14 atoms. Over many years, these atoms break down and so the proportion of C-14 in the material falls. By measuring the amount of C-14 in the remains of a living thing, scientists can figure out almost exactly how many years ago it was alive. This technique can be used for dating objects up to 50,000 years old.

▲ *A scientist takes a sample from a bone and then measures the proportion of carbon-14 to carbon-12 present in the sample. From this data the scientist can accurately estimate the age of the sample. The scientist wears gloves to prevent contamination with living tissue.*

Key Terms

- **Alpha particle:** the nucleus of a helium atom. This particle has two protons and two neutrons.
- **Isotope:** Atoms of a given element always have the same number of protons but can have different numbers of neutrons. These different versions of the same element are called isotopes.

different numbers of neutrons in its nuclei. The different versions of an element are called isotopes.

Hydrogen atoms, for example, exist as three isotopes. Most of them are simple hydrogen atoms with an atomic mass number of 1 (they have a single proton in the nucleus). About 0.015 percent of hydrogen atoms have a neutron as well as a proton in the nucleus. This isotope has an atomic mass number of two and is named deuterium, or heavy hydrogen. Finally, one in every billion trillion hydrogen atoms is the isotope tritium. This has two neutrons in the nucleus and so has an atomic mass number of 3.

DISPLAYING ISOTOPES

To ensure that people know what isotope an atom is, an atom's atomic mass number is always written above its symbol, and the atomic number is displayed below it. For example the main isotope of carbon is displayed as $^{12}_{6}\text{C}$. This isotope is also described as carbon-12 (C-12).

The atoms of both deuterium and tritium are radioactive. This means that

A Closer LOOK

Isotopes

Hydrogen exists as three isotopes. By far the most common isotope is ($_1^1$H). This has only a single proton in its nucleus and no neutrons. Deuterium ($_1^2$H) and tritium ($_1^3$H) are much rarer. They have one and two neutrons in their nucleus respectively.

Unlike the isotopes of hydrogen, most isotopes do not have special names and are therefore identified by their atomic number, such as carbon-12 ($_6^{12}$C) and carbon-14 ($_6^{14}$C).

The number of electrons does not change in different isotopes of the same element. The electron number remains the same as that of the atomic number (proton number).

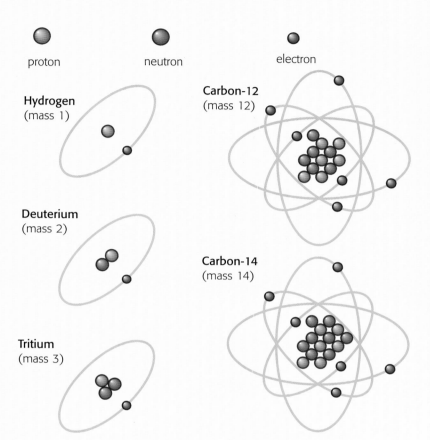

proton   neutron   electron

Hydrogen
(mass 1)

Deuterium
(mass 2)

Tritium
(mass 3)

Carbon-12
(mass 12)

Carbon-14
(mass 14)

their nuclei are unstable and they quickly break apart, releasing radiation (*see* pp. 56–65). Many of the less common isotopes of other elements are radioactive, too.

CALCULATING ATOMIC MASS

Chemists use a system of atomic masses based on the mass of each element that will react with a known mass of another. In effect, they are using an average mass based on the relative amount of each isotope of an element present. For example, most hydrogen atoms have an atomic mass number of 1, but a tiny amount of the element's atoms have atomic mass numbers of 2 and 3. Therefore the average atomic mass number for all hydrogen atoms is very slightly higher than 1 (1.00794 to be precise). This number is the element's atomic mass. To keep things simple, chemists often round down hydrogen's atomic mass to just 1.

RELATIVE ATOMIC MASS

Atoms are extremely light—a hydrogen atom weighs about 1.7 trillion trillionths of a gram. It would be very confusing to use units this small to compare one atom with another. Instead an atom's mass is expressed as a comparison with the masses of the atoms of other elements.

Chemists do this using atomic mass numbers. For example, hydrogen has an atomic mass of 1, helium has an atomic mass of 4, while carbon has an atomic mass of 12. This shows that carbon atoms are 12 times heavier than hydrogen atoms and three times heavier

▽ These blue crystals are made of molecules of copper sulfate ($CuSO_4$). The relative molecular mass (RMM) for copper sulfate can be calculated by adding the relative atomic masses (RAM) of each element. Copper (Cu) has a RAM of 64, sulfur (S) 32, and oxygen (O) 16. So, taking into account that $CuSO_4$ has four oxygen atoms, the RMM is $64 + 32 + (4 \times 16) = 160$.

Tools and Techniques

Mass spectrometer

The atoms of each element contain a certain set amount of matter. Chemists can use this fact to identify the ingredients of almost any material. They do this with a mass spectrometer. This is a machine that can detect the size of atoms and molecules inside a substance and calculate their proportions.

First the material is turned into a gas. The atoms and molecules in the gas are given a positive charge by ripping away some of their electrons using a strong electric current. The charged gas is formed into a beam by the injection magnet. The beam is then accelerated and fired through an analyzer magnet. This magnet makes the particles in the beam change direction. Lighter particles are deflected more than heavier ones. The particles then crash into a detector. Scientists can tell how heavy a particle is by where it hits the detector. Once they know how heavy the particle is they can figure out what type of atoms (or combination of atoms) it contains. They can then figure out what sort of substances were present in the sample.

▼ *This mass spectrometer is measuring how much of a sample is made of carbon-12, carbon-13, and carbon-14. The resulting data can be used for carbon dating (see box p. 30).*

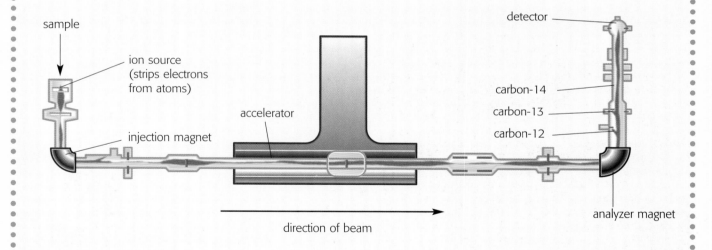

sample
ion source (strips electrons from atoms)
injection magnet
accelerator
detector
carbon-14
carbon-13
carbon-12
analyzer magnet
direction of beam

than helium atoms. Chemists often refer to atomic mass as relative atomic mass (RAM) because that name explains that the figure relates to the other elements.

Molecules contain more than one atom, and their mass is also important. This mass is measured as the total of the atomic masses of the atoms in the molecule. This adds up into the relative molecular mass (RMM). For example, a water molecule has two hydrogen atoms (atomic mass 1) and one oxygen atom (atomic mass 16). Therefore water's RMM is 18 (1+1+16). Knowing the RAMs and RMMs of different substances helps chemists figure out how those substances have combined and altered during chemical reactions.

TRY THIS

How many molecules?

Water has an RMM of 18. One mole of water has a mass of 18 grams (0.6 ounces). You can use this to calculate how many molecules a cup of water contains. Weigh an empty cup on a kitchen scale; using grams makes it easier. Pour some water into the cup and see how much heavier it is. The difference is the mass of the water. Divide this mass in grams by 18 to get the number of moles of water in the cup. Then multiply this figure by Avogadro's number to calculate the number of molecules. For example, if the water has a mass of 36 grams, then there are 2 moles that together contain 1,204,000,000,000,000,000,000,000 molecules, or 1.204 trillion trillion.

▲ If this beaker has a mass of 10 grams, then the water has a mass of 90 grams. This mass of water divided by 18 is 5. Therefore there must be 5 moles of water in the beaker.

MOLES

Chemists measure amounts of elements and compounds using a unit called a mole. The mole is defined as 12 grams (0.42 ounces) of carbon-12 atoms. (The RAM of this isotope is 12). Chemists chose this isotope to define the mole because carbon is a common element on Earth. One mole of any element has a mass equivalent to its atomic mass expressed in grams. For example, one mole of hydrogen has a mass of 1 gram (0.035 ounces), while one mole of lead, which has an atomic mass of 207, has a mass of 207 grams (7.3 ounces).

A mole of anything contains the same number of particles or pieces— 602,213,670,000,000,000,000,000. This number is more often written as 6.022×10^{23}, which means 6.022 multiplied by 10, 23 times. This is about 10 times the number of stars there are thought to be in the entire universe. This number is called Avogadro's number. It is named for the Italian scientist Amedeo Avogadro (1776–1856). Avogadro suggested that equal quantities of gases at a fixed temperature and pressure always contain the same number of atoms or molecules. Therefore a gallon of hydrogen gas contains the same number of atoms as a gallon of oxygen gas. However, the gallon of oxygen has a mass 16 times more than the gallon of hydrogen.

Using Avogadro's number, relative atomic mass numbers, and relative molecular mass numbers, chemists can easily figure out the number of atoms or molecules in a given sample.

Profile

Amedeo Avogadro

Amedeo Avogadro was born in Turin, Italy, on August 9, 1776. His father, Count Filippo Avogadro, was a lawyer, and Amedeo also trained as a lawyer but he became increasingly interested in physics and mathematics. In 1806, he was appointed demonstrator at the Academy of Turin. In 1811, he published a paper in which he suggested that equal volumes of gases at the same temperature and pressure contain the same number of molecules. This paper was to have a great influence on chemistry, although it was many years before the importance of his work was widely recognized. Avogadro's number is named in honor of his important contribution to the understanding of atoms and molecules, though he had no knowledge of this number or of the mole.

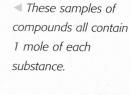
◀ These samples of compounds all contain 1 mole of each substance.

See ALSO ...
● Identifying Elements, Vol. 5: pp. 10–17.

Iron (III) chloride
$FeCl_3$, 161 g

Copper sulfate
$CuSO_4$, 160 g

Potassium iodide
KI, 166 g

Sodium chloride
NaCl, 58 g

Potassium permanganate
$KMnO_4$, 158 g

Cobalt nitrate
$Co(NO3)_2$, 183 g

4 Understanding Electrons

Electrons play an important role in chemical reactions. Their arrangement around an atom influences how easily an element will combine with other elements. They are also responsible for the production of light.

While most of an atom's matter is locked up in the dense nucleus at the center, most of its behavior is controlled by the tiny electrons that move around the nucleus. For example, electrons are the parts of the atom that are involved in chemical reactions.

ARRANGING ELECTRONS

Each electron in an atom has its own position. The electrons repel each other and never come into contact. They are also arranged in layers, called electron shells. Different atoms have a different number of shells, depending on how many electrons the atom has.

For example, hydrogen has just a single shell, containing its only electron. However, the atoms of uranium, the largest element, have their 92 electrons arranged in seven shells. The shell nearest the nucleus is the smallest. Shells get larger and can hold more electrons as they get farther away from the center of the atom.

The Sun releases particles that travel through space as the solar wind. When such a stream hits Earth, the particles in the solar wind collide with gas atoms in the atmosphere. The collisions knock the electrons in the atoms into another shell, emitting energy in the form of colored light as they go. These lights can be seen in the skies near the North and South poles and are called auroras.

A Closer LOOK

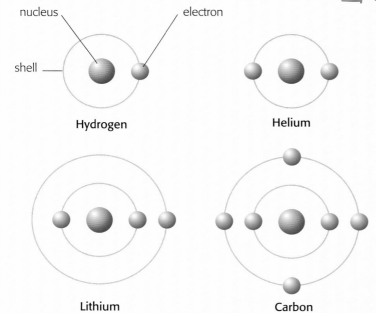

nucleus

electron

shell

Hydrogen

Helium

Lithium

Carbon

Electron shells

There is a strict order in which electrons fill the shells that surround the nucleus of an atom. Hydrogen, the first element in the periodic table (*see* vol. 5: pp. 4–9), has only one electron orbiting its nucleus. Helium, the second element, has two. Because the inner shell is very close to the nucleus, it can only hold two electrons. The next element is lithium, which has three electrons. Since the inner shell is full, the third electron has to orbit in the next shell. The second shell can hold a maximum of eight electrons. Atoms with full outer shells are stable and unreactive. Lithium needs seven electrons and carbon needs four electrons to be stable.

ENERGY LEVELS

Electron shells are also sometimes described as energy levels. The electrons in the shell nearest to the nucleus have the lowest amount of energy, and those farthest away have the highest. When an atom receives some energy, such as when it is heated, its electrons move to a higher energy level farther away from the nucleus. Atoms release energy when their electrons drop down to lower energy levels and move back toward the nucleus. This model of how an atom works was suggested by Niels Bohr

▶ *Niels Bohr was the Danish scientist who first suggested that electrons orbit the nuclei of atoms in shells. That idea was key to understanding atomic structure.*

▲ *The rays of light we see are caused by tiny particles called photons. Photons are created when the electrons that surround atoms change their energy level.*

(1885–1962), a Danish physicist, in 1913. It is still seen as one of the best ways to understand atoms.

HEAT AND LIGHT
Bohr's description of electron shells explains how light and other types of radiation are produced by atoms. Light is just one type of electromagnetic radiation. Others include radio waves, heat, ultraviolet light, and X-rays. All these types of radiation are produced in the same way, but some involve greater quantities of energy than others. Visible light lies in the middle of the spectrum—the name given to the range of radiation types. Some rays of light contain more energy than others. Our eyes see these differences as color. Blue light has more energy than yellow light, which has more than red light. Ultraviolet light (UV) and X-rays are two examples of radiation that have more energy than visible light. UV is the invisible radiation in sunlight that

Key Terms

- **Shell:** The orbit of an electron. Each shell can contain a specific number of electrons and no more.

- **Energy level:** The shells each represent a different energy level. Those closest to the nucleus have the lowest energy.

- **Electromagnetic radiation:** The energy emitted by a source, for example, X-rays, ultraviolet light, visible light, heat, or radio waves.
- **Photon:** A particle that carries a quantity of energy, usually in the form of light.

levels. As a result, when they drop down levels, they always release photons with exactly the same amount of energy. This amount of energy is called a quantum. Quanta (the plural of quantum) are fixed amounts of energy. It is not possible for an atom to release half a quantum. This fact forms the basis of quantum physics, the branch of science that investigates the forces that govern atoms.

causes sunburn and tanning. X-rays are used to take images of bones inside the body. Heat, also known as infrared radiation, contains less energy than light, as do radio waves.

RELEASING LIGHT

Most electromagnetic radiation is released from atoms. When an electron drops down an energy level, the atom releases a tiny particle called a photon. This particle is even smaller and lighter than an electron. Rays of light or other radiation are streams of photons being produced by atoms.

The photon carries the radiation's energy. The amount of energy it carries depends on how many energy levels the electron has dropped. If the electron has moved from a high energy level a long way from the nucleus to a shell near the nucleus, then the photon will carry high-energy radiation such as X-rays. Smaller falls release less energy.

FIXED AMOUNTS

The energy levels in an atom are fixed. They depend on the atom's size. Electrons can only move between energy levels. They cannot move halfway between two

A Closer LOOK

Photons

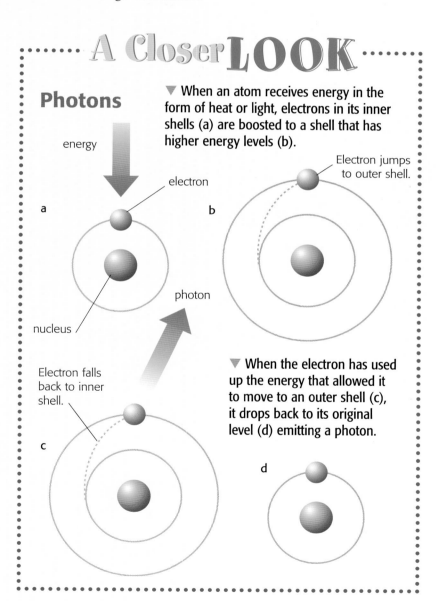

energy

electron

nucleus

a

b

Electron jumps to outer shell.

photon

Electron falls back to inner shell.

c

d

▼ When an atom receives energy in the form of heat or light, electrons in its inner shells (a) are boosted to a shell that has higher energy levels (b).

▼ When the electron has used up the energy that allowed it to move to an outer shell (c), it drops back to its original level (d) emitting a photon.

Chemistry in Action

Fireworks

Fireworks contain small amounts of explosives that blow up to produce a display of colors. The colors in fireworks are produced by certain chemicals mixed into the explosives. When the firework explodes, the atoms in these chemicals react with each other or the air and release energy as colored light. The color of the light depends on the elements in the firework. If a firework contains potassium compounds it will explode with a purple color. Lithium atoms produce a red color, while blue sparkles are caused by metals such as copper and cobalt.

▶ *The orange color of this firework indicates that it probably contains a sodium salt. Sodium always gives an orange flame when burned.*

Because atoms can only release a certain set of fixed amounts of energy, chemists can identify elements by the light they produce. Elements produce light and other radiation when they burn, or when they are heated. Potassium always produces a pale purple flame when it burns, while magnesium burns with a dazzling white flame. Each element produces a characteristic spectrum of colors that identify the element.

CHEMICAL BEHAVIOR

Electrons are the parts of an atom that take part in chemical reactions. When atoms join up, they lose, gain, or share their electrons with other atoms. The interaction of electrons creates the forces that hold the atoms together as molecules. Some elements are more likely to react and form compounds than others. Whether an element is reactive depends on how its electrons are arranged around the nucleus. The arrangement of electrons controls how easy it is for an atom to lose, gain, or share its electrons.

ELECTRON ARRANGEMENTS

The electrons in the outermost shell of an atom are the ones that take part in chemical reactions. Electron shells are most stable when they are completely full. The atoms of most elements have an outer shell that is only partially full. Lithium atoms have just one electron in their second shell, with seven of the spaces left empty. Chlorine has seven electrons in its outer shell, with room for one more.

Elements take part in chemical reactions that make their outer electron shells more stable. An atom can do this by giving away an electron, taking one from another atom, or it can share electrons with another atom so they both have full outer shells. For example, lithium atoms lose their single outer electron during reactions. As a result the atom also loses its whole second shell, and its first shell becomes the outer one. This shell has two electrons in it and so is full and very stable. On the other hand, chlorine atoms gain an electron when they react. This makes them stable by completing their own outer shell.

A Closer LOOK

Arranging electrons

This table shows the electron arrangement of some elements. Although the number of electrons in the inner shells varies as the atoms get larger, it is the number of electrons in the outer shell that governs how the atom reacts. Atoms with few outer electrons tend to lose them in reactions. Those with a nearly full outer shell form compounds by gaining or sharing electrons.

Element	Atomic number	Number of electrons in shell 1	shell 2	shell 3	Spaces in outer shell
Hydrogen	1	1			1
Helium	2	2			0
Lithium	3	2	1		7
Carbon	6	2	4		4
Magnesium	12	2	8	2	6
Chlorine	17	2	8	7	1

▲ *Lithium is an extremely reactive metal because it has one electron in its outer shell. It is even more reactive than hydrogen, which also has one electron in its outer shell. When added to water (H_2O) it will displace one of the hydrogen atoms to form lithium hydroxide.*

See Also ...
● *Chemical Bonds, Vol. 3: pp. 10–21.*
● *Atoms and Elements, Vol. 5: pp. 4–9.*

5 Reactions and Bonding

A chemical reaction changes one substance into another. During chemical reactions, atoms join with each other in new ways to form molecules. The molecules are held together by bonds between atoms.

A chemical reaction takes place when two or more substances are mixed together in the right conditions. The substances that are the ingredients needed for the reaction are called the reactants. During the reaction, the atoms inside the reactants separate from each other and reorganize. They form into one or more new substances. Chemists call these substances the products.

Reactants may be elements—with just one type of atom in them—or compounds that are made up of different types of atoms. The products may also be either elements or different compounds. During the reaction, no atoms are made or destroyed. All that happens is that they are rearranged. The number of atoms in the reactants is always the same as the number of atoms in the products.

One of the most important chemical reactions happens in plants. Plants use energy from the Sun to change water and carbon dioxide into sugars and oxygen.

Chemists show what has happened during a chemical reaction with a chemical equation. Chemical equations have two sides. The left-hand side shows the formulas of the reactants and the number of molecules that are needed for the reaction. The right-hand side shows the formulas of the products and the number of molecules that have been formed.

Coal burning in air is a simple reaction. Coal is mainly pure carbon. The carbon (C) reacts with molecules of oxygen (O_2) in the air. The product of this reaction is carbon dioxide gas (CO_2). The equation for this reaction is:

Carbon + oxygen → carbon dioxide

$$C \quad + \quad O_2 \quad \rightarrow \quad CO_2$$

Key Terms

Reactants: The ingredients necessary for a chemical reaction.

Products: The new substance or substances created by a chemical reaction.

BREAKING AND MAKING

The reaction between carbon and oxygen releases a lot of heat and light. Flames are super-hot carbon and oxygen atoms in the process of reacting. People have burned coal as a fuel for many centuries because the reaction it produces gives off so much heat.

Other reactions do not produce heat. Instead, these sorts of reactions need to be heated to make them work. When chalky calcium carbonate ($CaCO_3$) is

TRY THIS

Fizzing fun

You can see a quick, safe, and easy chemical reaction for yourself using two common household compounds. Mix a spoonful of baking soda (sodium bicarbonate) with vinegar (ethanoic acid). It is easier to see the results if you use a clear glass. When the baking soda is added it begins to fizz. The sodium bicarbonate reacts with the acid to produce three new compounds—water, sodium ethanoate, and carbon dioxide gas. The water and sodium ethanoate form a solution in the glass, while the gas, which causes the fizzing, bubbles away.

▶ Sodium bicarbonate and ethanoic acid use heat to react, so if you touch the glass during the reaction, it will feel a little cold.

Profile

A head for chemistry

French scientist Antoine Lavoisier (1743–1794) was one of the most important people in modern chemistry. He discovered the element oxygen, for example. However, Lavoisier also did a lot of work to show that atoms were not made or destroyed by chemical reactions but simply rearranged themselves to form new compounds. He carried out many experiments in which he carefully weighed the reactants and then all the products. Each time, his results showed that the amount of matter was always the same after a reaction as before it. The principle he proved is called the law of conservation of mass. Lavoisier was a member of a noble family from Paris. He inherited a large fortune from his parents but became even wealthier by collecting taxes from poor French workers. He used his large fortune to pay for his many experiments. After the French Revolution in 1789, his money was taken away. Lavoisier was eventually beheaded along with many other tax collectors.

◄ *Antoine Lavoisier was a political liberal who took part in events that led to the French Revolution. Nevertheless, he was beheaded for being a tax collector.*

heated it breaks up into calcium oxide (CaO) and carbon dioxide (CO_2). However, if the stone is left unheated, the reaction will not take place. The equation for this reaction is:

$$CaCO_3 \rightarrow CaO + CO_2$$

Whether a reaction produces or takes in heat depends on the bonds in the

◀ *Pouring the concrete for the Hoover dam had to be done very carefully to avoid cracking because of the huge amount of heat generated in the chemical reaction that produces concrete. Calcium compounds in cement react strongly with water to produce heat by the breaking and making of chemical bonds.*

reactants and products. These bonds hold the molecules together. During a chemical reaction, some of the bonds in the reactants are broken, and then new bonds form to make the products.

Energy is needed to break a bond, and energy is released when new bonds are formed. When a reaction has finished there is usually a difference between the energy used to break bonds and the energy released as new bonds are made. If more energy is released as the new bonds form than was used to break the old ones, then the reaction releases the spare energy as heat and light. If the new bonds release less energy than was used to break the old bonds, then the reaction needs extra energy to make it happen.

▲ *A lime kiln is used to heat chalk, a form of calcium carbonate, to produce a chemical reaction. Heat breaks the bonds in calcium carbonate, turning it into calcium oxide (lime) and carbon dioxide. Lime can be used as a fertilizer and was once used to paint buildings white.*

shell and so strongly attracts electrons from other atoms. That makes fluorine very likely to react.

Elements with very low electronegativity are also highly reactive. These elements have just a few electrons in their outer shell. Elements like this are generally metals. Metals do not pull electrons toward them and they hold on to their outer electrons very lightly. This makes them electropositive. The most

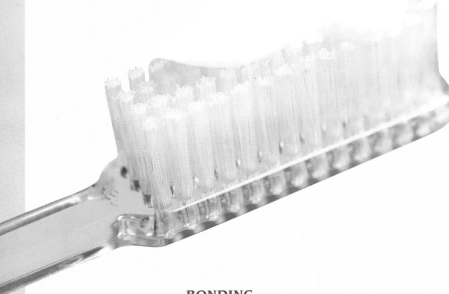

▲ *Toothpaste commonly contains compounds of fluorine, such as sodium fluoride, which help prevent tooth decay. Fluorine is the most electronegative element and so is highly reactive. In its elemental form it is a poisonous gas. When combined with other elements, however, it may have very different properties.*

BONDING

Atoms can bond in a number of ways. There are three main types of bonds— ionic, covalent, and metallic (*see* vol. 3: pp. 10–21). The way particular elements form bonds depends on how many electrons their atoms have in their outer shell. Some elements produce stronger bonds than others. Strong bonds need a lot of energy to break them. They are produced when two very reactive atoms bond.

An element's reactivity depends on something called electronegativity. That is a measure of how strongly one atom holds on to its electrons and pulls on the electrons of other atoms. Elements that have an outer electron shell with only a few empty spaces left to fill will pull on electrons the most strongly. Nonmetals are the most electronegative. Fluorine gas is the most electronegative element. It has just one space to fill in its outer

Key Terms

Electronegativity is the power of an atom to attract an electron. Nonmetals, which have only a few spaces in their outer shell, are the most electronegative. Metals, which have several empty spaces, are the least electronegative elements. These metals tend to lose electrons in chemical reactions. Metals of this type are termed electropositive.

Fluorine   most electronegative
Oxygen
Chlorine
Nitrogen
Bromine
Iodine
Sulfur
↑

Sodium
Radium
Barium
Potassium
Rubidium
Cesium
Francium   most electropositive
↓

electropositive elements, such as cesium and francium, have only one electron in their outer shell. When they lose this electron, they become stable.

IONIC BONDS

When electronegative and electropositive atoms react they form ionic bonds. An ion is a charged version of an atom. A positively charged ion is an atom that has lost one or more of its electrons. Atoms that gain electrons become negative ions. The size of an ion's charge depends on how many electrons it has lost or gained. For example, chlorine (Cl) atoms gain a single electron to become negative (Cl^-) ions. However, calcium atoms can lose two electrons to become positive (Ca^{2+}) ions.

Charged objects are attracted to objects with an opposite charge. This is the same force that keeps an electron orbiting a nucleus. Ions with opposite charges are pulled toward each other by this force. It is this attraction that holds the two of them together, forming an ionic bond.

Chemistry in Action

Ionic bond

Common salt (sodium chloride) is a compound held together by an ionic bond. When an ionic bond forms, the single electron in a sodium atom's outer shell breaks free, leaving behind a full outer shell. The sodium atom becomes a positive sodium ion (Na^+). The free electron moves to the chlorine atom. It takes the final place in that atom's outer shell, making the atom a negative chloride ion (Cl^-). The two ions have opposite charges so are attracted to each other. This is the force that bonds them together.

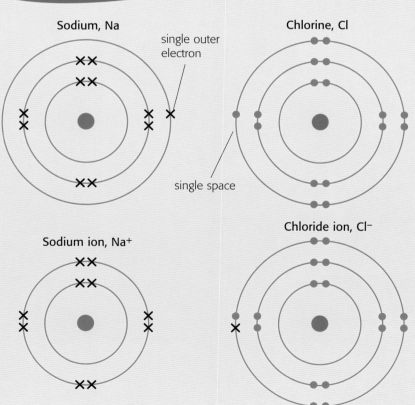

Sodium, Na

single outer electron

single space

Chlorine, Cl

Sodium ion, Na⁺

Chloride ion, Cl⁻

▼ *The atoms in a rubber band are joined in long coiled molecules. These molecules are connected to each other by covalent bonds. When the rubber band is stretched, the long molecules uncoil until the covalent bonds prevent further stretching. Any further tension on the band causes these covalent bonds to break and the band snaps.*

SHARING ELECTRONS

Some elements are neither particularly electropositive nor electronegative. That is because the outer shell is more or less half filled. Such elements have just as many electrons to lose as they have to gain to have a full outer shell. For example, carbon atoms have four electrons in their outer shell. To become stable by having a full outer shell, a carbon atom has two options. It can

either pull four electrons off other atoms or it can release its four outer electrons. Both are very unlikely because they would require huge amounts of energy. Instead carbon and similar elements get a full outer shell by sharing electrons. The shared electrons sit in the outer shells of both atoms. That is called a covalent bond. Each of the shared electrons is being pulled by the nucleus of both atoms. This is the force that holds the atoms together. Each covalent bond involves two electrons being shared. Some atoms can form more than one covalent bond at a time. For example, a carbon atom can form four covalent bonds at once. These four bonds can be with four other atoms. Sometimes, however, two atoms share two pairs of electrons. These are called double bonds. Carbon can even form triple bonds, which share three pairs of electrons.

The Eiffel Tower, once the world's tallest structure, is made from iron. Iron is strong enough to support such a tall structure because of its close metallic bonds.

METALLIC BONDS

Metals are generally hard solids that can be bent or stretched without breaking. They are also good conductors (carriers) of heat and electricity. These properties are the result of the way in which a metal's atoms are bonded.

The connections between metals' atoms are called metallic bonds. Metallic bonds

Chemistry in Action

Sharing electrons

Nonmetals tend to form covalent bonds with each other. Many nonmetal atoms combine with atoms of the same element to form simple molecules using these bonds. Oxygen, for example forms O_2 and fluorine forms F_2. Carbon forms bonds with hydrogen to form methane. By sharing electrons in a covalent bond, the elements form molecules that are more stable than atoms on their own.

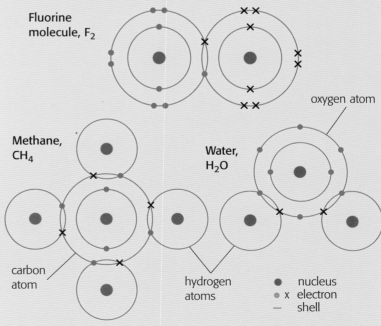

Fluorine molecule, F_2

oxygen atom

Methane, CH_4

Water, H_2O

carbon atom

hydrogen atoms

- nucleus
- x electron
- — shell

involve metal atoms sharing some or all of their outer electrons. Nearly all metallic elements have only one or two outer electrons. Only a few metals, such as lead, bismuth, and tin, have more. When a metal's atoms are packed closely together, as in a solid or liquid, the outer electrons from each atom break free. The

▲ Copper is often used in electrical wiring because it is a good conductor of electricity. The electrical current is carried by the "sea" of electrons that surround the atomic nuclei.

Electron sea

Within a metal, nuclei are attracted in all directions to the "sea" of electrons that surround them, so the nuclei and electrons are held together very strongly.

Because the sea of electrons can move, it carries heat and electricity well. (Electricity is a flow of electrons moving through a substance.) That is why most metals are very good conductors (carriers) of heat and electricity.

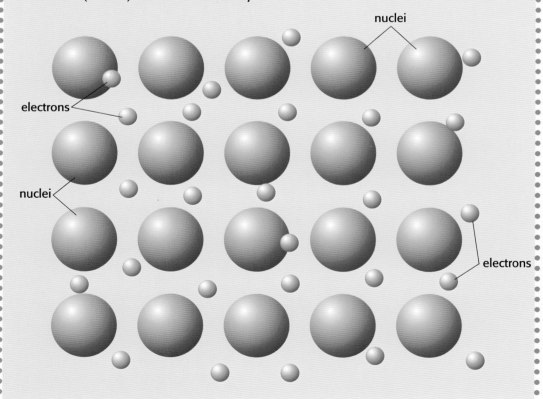

free electrons form a "sea" of electrons that can move around and are shared by all the atoms. The positively charged nucleus inside a metal atom is attracted to the negatively charged electron sea all around it. This force holds metal atoms in place.

INTERMOLECULAR BONDS

Ionic, covalent, and metallic bonds hold atoms together. However, there are other forces that make atoms and molecules cling together.

Most of these are very weak. For example, tiny temporary forces are produced by the random movement of the electrons. The electrons in an atom or molecule are generally spread out evenly. However, they are all constantly moving and by pure chance they can sometimes all gather in one place at the same time. This gives one end of an atom or molecule a negative charge and the other end becomes positive. These charges exist for only a very short time but they have an effect by pushing and pulling on the atoms around them.

These forces are named van der Waals forces for Dutch physicist Johannes van der Waals (1837–1923). He

▲ *A gecko is a lizard that can cling to surfaces using van der Waals forces. These forces occur between a surface and tiny hairs on the gecko's feet.*

was the first to realize the importance of these forces and how they affect the behavior of gases and liquids.

Larger molecules produce stronger van der Waals forces than smaller ones. This increases the melting and boiling points of the larger molecules. Even though they are tiny, the van der Waals forces hold molecules together and make it harder to break the bonds between them.

DIPOLE ATTRACTIONS

Some types of molecules always have charged ends. These charged regions are called dipoles. Dipoles are produced when one atom in a molecule is more electronegative than the others. As a result all the outer electrons in the molecule are drawn toward that atom. Because more of the electrons are at one end, that end, or pole, is negative. The other pole is positive. The charged poles are attracted to the oppositely charged pole of a nearby molecule. Dipoles attract the molecules to one another and make them arrange themselves in a repeating pattern, where the oppositely charged poles are next to each other.

HYDROGEN BONDS

Hydrogen bonds are an example of a dipole attraction. As their name suggests, these bonds always involve an atom of hydrogen. When hydrogen is bonded to highly electronegative elements, such as fluorine, it often forms a positive pole. The hydrogen atom's only electron is pulled away by the other atom, leaving just the hydrogen's nucleus. This nucleus is a single proton and has a strong positive charge.

Water is an example of a compound that produces hydrogen bonds. The oxygen atom in the water molecule pulls the electrons from the hydrogen atoms. The oxygen has a slightly negative charge, and each hydrogen is slightly positive. The positive hydrogen atoms are attracted to the negative pole of another water molecule.

The hydrogen bonds in water ensure that it is a liquid in the normal

▼ Water (H$_2$O) molecules form hydrogen bonds. The hydrogen atoms in water have a slight positive charge and the oxygen atoms have a slight negative charge, owing to the arrangement of electrons around each nucleus. These opposite charges attract each other and hold the water molecules loosely together.

Key Terms

- **Dipole attraction**: The attractive force between charged ends of molecules.
- **Hydrogen bond:** Dipole attraction that always involves a hydrogen atom.
- **Intermolecular bonds:** The bonds that hold molecules together. These bonds are weaker than those between atoms in a molecule.
- **Van der Waals forces:** Short-lived forces between atoms and molecules.

conditions found on Earth's surface. Without these bonds, water molecules would not be so strongly bonded to each other. As a result, the boiling point of water would be a much lower temperature, and water would be a gas in normal conditions.

SHAPES OF MOLECULES

As well as sometimes creating weak forces between molecules, the position of electrons in a molecule has an effect on its shape. Opposite charges attract and like charges repel each other.

The electrons in a molecule repel each other and they tend to stay as far away from each other as possible. Electrons in an atom's outer shell will repel each other with equal force. However, an electron that is being shared with another atom to form a bond cannot push the other electrons away as strongly. As a result, pairs of shared electrons are often pushed away from the other, unbonded electrons.

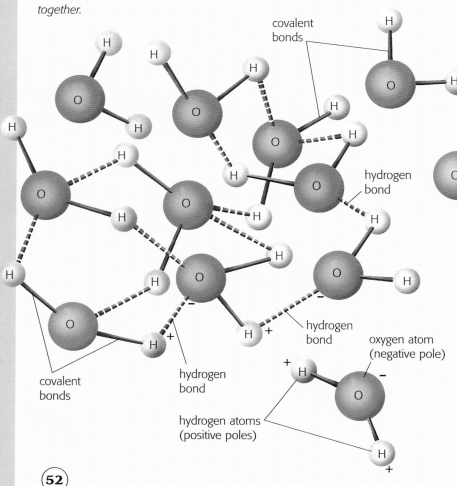

covalent bonds

hydrogen bond

hydrogen bond

oxygen atom (negative pole)

hydrogen atoms (positive poles)

hydrogen bond

covalent bonds

Chemistry in Action

Floating ice

Water is an unusual substance. Most substances contract a little when they freeze. But water expands when it turns to ice. As a result ice is less dense than water. This ensures that ponds and rivers always freeze from the top down and makes it possible for huge icebergs to float in the ocean. Ice takes up more room than water because of hydrogen bonds. As water freezes, these bonds force the molecules into a widely spaced crystal structure. When ice melts, hydrogen bonds have less of an effect, the bonds continually break and re-form, and the molecules mingle more closely with each other, taking up less room.

The uneven distribution of outer electrons has an effect on the shape of a molecule. A molecule with two atoms, such as a chlorine gas molecule (Cl_2), always forms a straight line and can be thought of as a miniature dumbbell. However, when a molecule has more than one bond in it, the shape can be more complicated.

A molecule of methane gas contains one carbon atom at its center. There are four hydrogen atoms bonded to the carbon. Carbon has four electrons, and each one makes a covalent bond with one of the hydrogens. Therefore all pairs of electrons are identical and they all repel each other equally. The result of this is that methane molecules form a tetrahedron—a type of pyramid.

Water molecules, however, are shaped by the force of unbonded electrons. Each molecule is made up of two hydrogen atoms joined to a single oxygen atom by covalent bonds. The molecule is not straight but bent, with both the hydrogen atoms being on the same side of the oxygen atom. That is because the two pairs of shared electrons in the molecule's bonds are being repelled by the oxygen atom's six other electrons.

Atoms can also form molecules that have complex shapes. Carbon atoms can join to make spherical molecules of connected hexagons and pentagons (*see* vol. 8: pp. 8–17). These molecules are called buckminsterfullerenes for the American engineer and inventor Buckminster Fuller (1895–1983), who designed domes that have the same structure as these carbon molecules. Carbon can also form sheets of hexagons that roll up to form hollow tubes.

▼ *Around hot springs there is often a smell of rotten eggs. This is caused by hydrogen sulfide (H_2S), which has a molecular shape similar to that of water.*

See Also ...

Energy in Chemical Reactions, Vol. 4: pp. 4–17.

Properties of Metals, Vol. 6: pp. 4–13.

A Closer LOOK

Molecule shapes

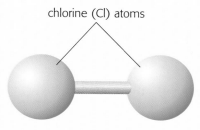
chlorine (Cl) atoms

A chlorine gas molecule (Cl_2) forms a dumbbell shape.

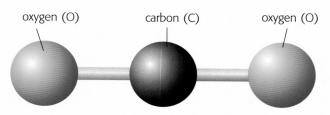
oxygen (O)　　carbon (C)　　oxygen (O)

In carbon dioxide (CO_2) the atoms form a straight line.

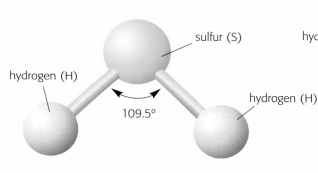
sulfur (S)

hydrogen (H)

hydrogen (H)

109.5°

Hydrogen sulfide (H_2S) has an angle of 109.5° between the hydrogen and sulfur atoms. Hydrogen sulfide has a similar molecular shape to water (H_2O), which has a bond angle of 104°.

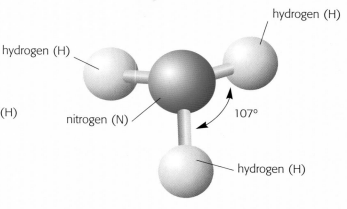
hydrogen (H)

hydrogen (H)

nitrogen (N)

107°

hydrogen (H)

An ammonia (NH_3) molecule has the shape of a flattened pyramid. The angle between any two of the hydrogen atoms and the nitrogen atom is 107°.

Methane (CH_4) forms a pyramid with a single carbon atom at its center. The angle between any two hydrogen atoms and the carbon atom is 109.5°.

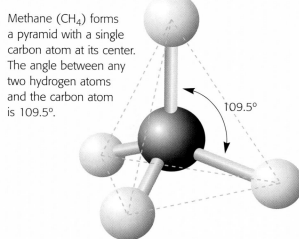
109.5°

A buckminsterfullerene (C_{60}) has 60 carbon atoms arranged in a sphere made of pentagons and hexagons (five- and six-sided shapes).

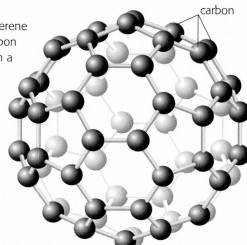
carbon

6 Radioactivity

Some elements are unstable. Their atoms break apart and release energy in the form of radiation. This process is called radioactivity. Radioactive elements are dangerous but they also have many uses.

Atom bombs make use of the instability of radioactive elements. By exploding these elements, huge quantities of energy are released, along with invisible but deadly radiation.

An atom is described as radioactive if it has an unstable nucleus. An atom's nucleus is made of positively charged protons and neutrally charged neutrons. Particles with the same charge repel each other, and that is true of protons. However, the protons stay together in the nucleus instead of being forced apart. That happens because there is an even stronger attractive force that bonds the protons and neutrons in place. Radioactivity occurs when this strong force cannot keep the nucleus together.

Inside the nucleus of a radioactive atom, the ratio of protons to neutrons makes it hard for the strong force to hold the particles together. Eventually small amounts of the nucleus break off and escape from the atom. This process is called radioactive decay. Radioactive decay is type of nuclear reaction. Nuclear reactions are different from chemical reactions. Chemical reactions involve just the electrons in an atom. Nuclear reactions result in changes to the nucleus of an atom.

RADIOACTIVE ELEMENTS

Radioactive elements are usually those with very large atoms. Their atoms have so many particles in each nucleus that they are more unstable than smaller atoms. For example, uranium atoms have between 234 and 238 particles in their nucleus. Uranium is one of the most common radioactive elements. There are nine other naturally occurring elements with atoms that are always radioactive: bismuth, polonium, astatine, radon, francium, radium, actinium, thorium, and protactinium.

▶ *These crystals contain uranium, one of the most common radioactive elements. The crystals are processed to produce uranium metal pellets, which are used to fuel nuclear power plants.*

However, some isotopes of other elements are radioactive. Isotopes are atoms that have the same atomic number but have a different atomic mass. One of the rare hydrogen isotopes—tritium—is radioactive. Radioactive isotopes are known as radioisotopes. Radioisotopes are always less common than the stable version of an element. For example, one in every trillion carbon atoms is the radioisotope carbon-14. All the others are stable carbon-12.

TYPES OF RADIATION

Radioactivity produces three types of radiation (*see* box left). These are alpha particles, beta particles, and gamma rays. (Alpha, beta, and gamma are the first three letters in the Greek alphabet.) Most nuclear reactions release either an alpha particle or a beta particle. All nuclear reactions produce gamma rays.

A Closer LOOK

Radiation types

beta particle (electron)

alpha particle (helium nucleus)

gamma ray

radioactive atomic nucleus

Alpha particles are two protons and two neutrons stuck together. This is the same as a nucleus of a helium atom, so alpha particles are often written as $^4_2$He. The 4 refers to the atomic mass of the nucleus (how heavy it is) and the 2 refers to the atomic number (the number of protons) of the atom. Because alpha particles do not have any electrons attached to them, the protons they contain give them a positive charge.

Most beta particles are fast-moving electrons. They have a negative charge just like the electrons that orbit atoms (*see* pp. 36–41). Beta particles are produced when a neutron in an unstable nucleus breaks down into a proton. A proton is slightly smaller than a neutron, and the leftover matter flies away in the form of an electron.

Gamma rays are energy waves that belong to the electromagnetic spectrum.

▼ People who work with radioactive substances put up signs to warn of the dangers because radiation cannot be seen, smelled, or felt. This sign shows how much radiation a human would receive if he or she stayed in the room for one hour.

Chemistry in Action

Antimatter

A few beta particles have a positive charge. They are the same size as electrons but have an equal and opposite charge. Particles like this are called positrons. Scientists describe particles such as a positron as antimatter. It is exactly the same as a particle of matter—in this case an electron—but has an opposite charge. When matter and antimatter particles meet they are destroyed completely, releasing gamma rays. Positron beta particles are produced when a proton in a nucleus turns into a neutron.

This spectrum also includes light, heat, radio waves, and X-rays. However, gamma rays contain more energy than any other type of wave. Some nuclear reactions also produce X-rays.

DANGERS OF RADIATION

All radiation produced by radioactive substances is dangerous. Alpha and beta particles are charged, and can rip electrons away from other molecules. This process is called ionization.

If radiation particles get into a person's body, they can damage the complex molecules inside cells. As a result, important cells may die or go wrong in other ways. For example, sometimes cells go wrong by growing in an uncontrolled way. This unusual growth produces a tumor in the body, which is a type of cancer.

Alpha particles are the largest type of radiation and cause the most damage. They are easy to stop, however, because they cannot pass easily through solid objects. They can be blocked by a sheet of paper or clothing.

Beta particles are much smaller than alpha particles and so can penetrate farther into solid objects. However, once they are inside the body, they do less damage than alpha particles because they are so small. Beta particles can be blocked by a thin sheet of metal.

Gamma rays cause ionization inside the body. They are more penetrating than alpha particles or beta particles. Gamma rays can pass through clothing, metal sheets, and most other everyday objects. Only thick slabs of lead will stop them completely. However, only a fraction of

Key Terms

- **Electromagnetic spectrum:** The range of energy waves that includes light, heat, and radio waves.
- **Ionization:** The formation of ions by adding or removing electrons from atoms.
- **Isotopes:** Atoms with the same atomic number but a different atomic mass.
- **Radiation:** The products of radioactivity —alpha and beta particles and gamma rays.

the gamma rays are absorbed by body tissues and many pass right through the body without having any effect.

CHANGING ELEMENTS

After a nucleus has decayed, the number of protons it contains changes. If the nuclear reaction released an alpha particle, there are two fewer protons in

▼ Radioactive particles have different powers of penetration. An alpha particle is relatively large and is easily stopped by a sheet of paper. Beta particles are smaller and faster. It takes a sheet of aluminum 1/5-inch (5 mm) thick to stop them. Gamma rays are very difficult to stop—a sheet of lead 4/5-inch (2 cm) thick is the minimum needed.

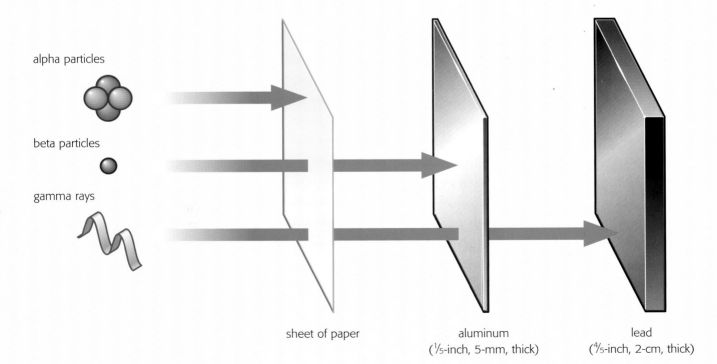

alpha particles

beta particles

gamma rays

sheet of paper

aluminum
(1/5-inch, 5-mm, thick)

lead
(4/5-inch, 2-cm, thick)

◀ Radon is a natural radioactive gas that collects under houses in some areas. Testing kits are available to detect its presence.

reactions before an atom decays into a stable element and all radioactive behavior comes to an end. Several different elements are produced by the string of nuclear reactions. This is called the decay chain. For example, the decay

the nucleus. If the reaction released a beta particle, one neutron changed into a proton. The nucleus now has one more proton than before. In both cases, the reactions change the atom's atomic number and it becomes a new element.

For example, an atom of the most common uranium isotope, U-238, which has the atomic number 92, releases an alpha particle as it decays. It loses two protons and turns into an atom of thorium. That element has the atomic number 90. Thorium is also radioactive. When the thorium atom decays it releases a beta particle. As a result of that type of decay, the atom's nucleus loses a neutron but gains a proton. That gives it a new atomic number of 91, and the atom has become protactinium.

DECAY CHAIN

In the above example of radioactive decay, one radioactive atom decays into another. It may take many nuclear

A Closer LOOK

Nuclear transformation

Radioactive elements can undergo any of three types of decay. These types are alpha particles, beta particles, and gamma rays. Often the type of decay varies with each stage of the process. The atomic number and mass number of the element change with alpha and beta decay, as these involve the loss of protons or neutrons, but stay the same when gamma rays are given off.

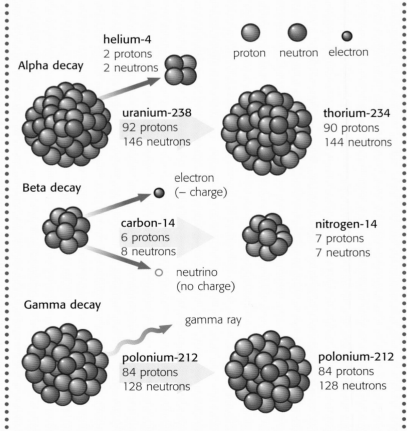

Uranium-238 decay chain

Radiation emitted alpha (α) or beta (β)		Isotope	Half-life
α		uranium-238	4.5 billion years (4.5×10^9)
β		thorium-234	24.5 days
β		protactinium-234	1.14 minutes
α		uranium-234	233,000 years (2.33×10^5)
α		thorium-230	83,000 years (8.3×10^4)
α		radium-226	1,590 years
α		radon-222	3.825 days
α		polonium-218	3.05 minutes
β		lead-214	26.8 minutes
β		bismuth-214	19.7 minutes
α		polonium-214	0.00015 seconds (1.5×10^{-4})
β		lead-210	22 years
β		bismuth-210	5 days
α		polonium-210	140 days
		lead-206	stable

▲ *The radioactive decay chain for uranium-238 takes 14 steps before one atom of uranium is finally converted into a stable isotope of lead.*

chain of uranium-238 contains a total of 14 other isotopes. It eventually ends when stable lead-206 atoms are produced (*see* box left).

The most common naturally occurring radioactive elements on Earth are thorium and uranium. They are found in rocks throughout the world. Most of the other radioactive substances that occur are produced as part of the decay chains of more common elements.

The rarer radioactive elements include radon, the only radioactive element that is a gas, and francium, the most reactive (and rarest) of all metals. The rare elements are more unstable than thorium and uranium and break down quickly.

HALF-LIVES

The rate at which a radioactive element decays is called its half-life. This is the time it takes for half of a sample of radioactive elements to decay. Imagine an element with a half-life of one year. Let us begin with 800 atoms of this element. After the first year only 400 are left. Over the next year, half of the atoms decay, leaving 200. After the third year, 100 atoms remain. This continues until there are no atoms of the element left (*see* graph, p. 63).

Common radioactive isotopes are relatively stable and have a very long half-life. For example, thorium-232 has a half-life of 14 billion years. Uranium-238 is less stable but has a half-life of 4.5 billion years. Bismuth-209 was only recently discovered to be a radioactive isotope. That is because bismuth-209 decays so slowly. Its half-life is 19 billion billion years.

Profile

The Curies

Polish physicist Marie Curie (1867–1934) and her French husband Pierre (1859–1906) were pioneers in the study of radioactivity. They even coined the term *radioactivity* to describe what they were investigating. Although radiation in the form of X-rays was already known by the time the Curies began work in 1898, nobody really knew where it came from. Henri Bequerel (1852–1908) had already shown that uranium ores gave off radiation. The Curies discovered that the strength of radiation emitted by a uranium ore depended on the amount of uranium atoms in the compound. The Curies also noticed that a mineral called pitchblende contained uranium compounds that produced more radiation than they expected. They realized that the rock must also contain other radioactive elements. They managed to identify two of these. The two new elements were named polonium (for Marie's home country Poland) and radium. The Curies received the 1903 Nobel Prize for physics for their work. Marie Curie also won the 1911 Nobel Prize for chemistry for her discovery of radium and polonium.

Unfortunately the harmful effects of radioactivity were not known at the time of the Curie's investigations, and Marie Curie died of leukemia caused by her exposure to radiation. Her notebooks are still so radioactive that they cannot be handled. Pierre was often burned by radiation and, like his wife, suffered from poor health but died in a street accident before the effects of radiation could kill him.

◄ *This illustration from a French magazine shows Marie and Pierre Curie at work in their laboratory. Their discovery of two new elements brought them international fame and attracted wide interest in their work.*

A Closer LOOK

Decay curve

All radioactive elements decay at a constant rate called a half-life. The half-life of each radioactive element is different. Whatever the period of the half-life is, the same curve results when plotted on a graph. The quantity of the element reduces by half by the end of each half-life, until it reaches zero and the element has become stable.

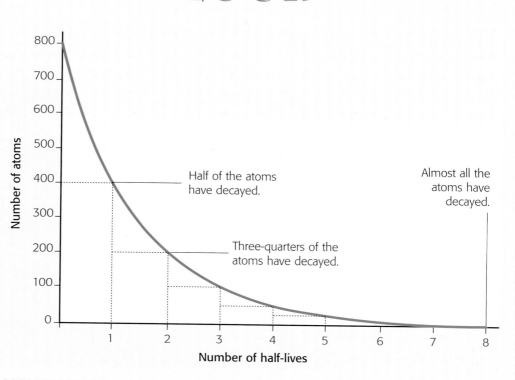

Half of the atoms have decayed.

Almost all the atoms have decayed.

Three-quarters of the atoms have decayed.

Number of atoms

Number of half-lives

The most unstable elements have the shortest half-life. They are extremely rare because they do not exist for long. Even the most stable isotope of francium has a half-life of just 22 minutes. Chemists think that the total amount of francium on Earth adds up to less than 1 ounce (28 grams) at any one point in time. Some isotopes are even more unstable and have a half-life of millionths of a second. These isotopes keep undergoing radioactive decay until they reach a state in which they become stable.

SPLITTING ATOMS

Radioactive decay is not the only type of nuclear reaction. For example, a process called nuclear fission is what powers nuclear power plants and causes nuclear bombs to explode. Another phrase for fission is "splitting the atom." Fission occurs when a neutron is fired at certain radioactive isotopes, generally uranium-235. The neutron causes the uranium atom to split in two, producing two smaller atoms and a great deal of heat. Each fission also releases more neutrons, which can cause further fissions in neighboring U-235 atoms. Nuclear reactors are used to keep this chain-reaction process under control. However, in nuclear bombs the chain reaction is allowed to occur at full speed. Huge amounts of energy are released in a devastating explosion.

▲ A graph showing the decay of a radioactive element takes the shape of a decreasing curve. The example above shows how 800 atoms of an element with a half-life of one year reduces in quantity. At the end of the first year, 400 atoms are left. After eight years (eight half-lives), only about one atom of the original radioactive element remains.

▼ *Supernovas such as this are the birthplace of the heaviest elements. As giant stars run out of fuel, they produce heavier atoms, which are then flung into the wider universe when the star explodes.*

THE ORIGIN OF MATTER

Another type of nuclear reaction is called fusion. This is the opposite of fission. Fusion involves two small atoms being squeezed together to form a larger one. It takes a large force to fuse atoms together, but when they fuse a huge amount of energy is also released.

Fusion is the reaction that powers the Sun. The Sun is a ball of hot gas. At its center, atoms are being pushed together with such force that they fuse. In most cases hydrogen atoms fuse to make helium atoms. This process produces the Sun's heat and light.

Fusion is the process by which the many different elements found in nature were created. In the largest stars, where there is enough force pushing into the center, larger atoms are formed by the fusion process. Atoms up to the size of iron are produced in this way.

The fusion reactions needed to produce the heaviest elements, such as gold and

Chemistry in Action

Artificial elements

There are 92 naturally occurring elements, 11 of which are radioactive. In addition, some elements, such as carbon, have radioactive isotopes. Scientists have also made artificial elements, all of which are radioactive. They do this by bombarding large natural elements with smaller ions. The atoms are bombarded at such high speed that the ions merge with them into huge artificial atoms.

Most of the artificial elements are heavier than uranium. Twenty-six artificial elements have been made so far. Only 20 of these have been given names. Many have been named for famous scientists and include curium (atomic number 96), einsteinium (99), and bohrium (107). Seaborgium (106) was named for American Glenn Seaborg (1912–1999). Seaborg helped make a number of new elements, including plutonium (94), americium (95), berkelium (97), californium (98), and mendelevium (101). These elements all have a very short half-life. For example, the half-life of bohrium is just 10 seconds.

▶ *Nitrogen-13 is an artificial radioactive isotope with a half-life of 10 minutes. It has to be made just before it is used in a hospital scanner.*

uranium, require gargantuan amounts of energy. Fusion reactions such as these are only possible inside huge explosions called supernovas. Supernovas are caused by the most giant stars running out of the fuel they need to keep shining. The pressure inside a supernova is so high that it can cause even heavy atoms to fuse. The atoms fused inside the explosion are flung into space, where they form clouds of dust and gas. Over billions of years the dust gradually clumps together to form planets. Earth was formed by this process about 4.6 billion years ago. Everything we see around us was once part of a star.

See Also ...
• Nuclear Reactions, Vol. 3: pp. 58–65.
• Rare-earth Metals, Vol. 5: pp. 60–65.

More Information

BOOKS

Atkins, P. W. *The Periodic Kingdom: A Journey into the Land of Chemical Elements*. New York, NY: Basic Books, 1997.

Bendick, J., and Wiker, B. *The Mystery of the Periodic Table (Living History Library)*. Bathgate, ND: Bethlehem Books, 2003.

Berg, J., Stryer, L., and Tymoczko, J. *Biochemistry*. New York, NY: W. H. Freeman, 2002.

Brown, T., Burdge, J., Bursten, B., and LeMay, E. *Chemistry: The Central Science*. 10th ed. Englewood Cliffs, NJ: Prentice Hall, 2005.

Cobb, C., and Fetterolf, M. L. *The Joy of Chemistry: The Amazing Science of Familiar Things*. Amherst, NY: Prometheus Books, 2005.

Cox, M., and Nelson, D. *Lehninger's Principles of Biochemistry*. 4th ed. New York, NY: W. H. Freeman, 2004.

Davis, M. *Modern Chemistry*. New York, NY: Henry Holt, 2000.

Herr, N., and Cunningham, J. *Hands-on Chemistry Activities with Real Life Applications*. Hoboken, NJ: Jossey-Bass, 2002.

Houck, Clifford C., and Post, Richard. *Chemistry: Concepts and Problems*. Hoboken, NJ: Wiley, 1996.

Karukstis, K. K., and Van Hecke, G. R. *Chemistry Connections: The Chemical Basis of Everyday Phenomena*. Burlington, MA: Academic Press, 2003.

LeMay, E. *Chemistry: Connections to Our Changing World*. New York, NY: Prentice Hall (Pearson Education), 2000.

Oxlade, C. *Elements and Compounds*. Chicago, IL: Heinemann, 2002.

Poynter, M. *Marie Curie: Discoverer of Radium **(Great Minds of Science)***. Berkeley Heights, NJ: Enslow Publishers, 2007.

Saunders, N. *Fluorine and the Halogens*. Chicago, IL: Heinemann Library, 2005.

Shevick, E., and Wheeler, R. *Great Scientists in Action: Early Life, Discoveries, and Experiments*. Carthage, IL: Teaching and Learning Company, 2004.

Stwertka, A. *A Guide to the Elements*. New York, NY: Oxford University Press, 2002.

Tiner, J. H. *Exploring the World of Chemistry: From Ancient Metals to High-Speed Computers*. Green Forest, AZ: Master Books, 2000.

Trombley, L., and Williams, F. *Mastering the Periodic Table: 50 Activities on the Elements*. Portland, ME: Walch, 2002.

Walker, P., and Wood, E. *Crime Scene Investigations: Real-life Science Labs for Grades 6–12*. Hoboken, NJ: Jossey-Bass, 2002.

Wertheim, J. *Illustrated Dictionary of Chemistry* (Usborne Illustrated Dictionaries). Tulsa, OK: Usborne Publishing, 2000.

Wilbraham, A., et al. *Chemistry*. New York, NY: Prentice Hall (Pearson Education), 2000.

Woodford, C., and Clowes, M. *Routes of Science: Atoms and Molecules*. San Diego, CA: Blackbirch Press, 2004.

WEB SITES

The Art and Science of Bubbles
www.sdahq.org/sdakids/bubbles
Information and activities about bubbles.

Chemical Achievers
www.chemheritage.org/classroom/chemach/index.html
Biographical details about leading chemists and their discoveries.

The Chemistry of Batteries
www.science.uwaterloo.ca/~cchieh/cact/c123/battery.html
Explanation of how batteries work.

The Chemistry of Chilli Peppers
www.chemsoc.org/exemplarchem/entries/mbellringer
Fun site giving information on the chemistry of chilli peppers.

The Chemistry of Fireworks
library.thinkquest.org/15384/chem/chem.htm
Information on the chemical reactions that occur when a firework explodes.

The Chemistry of Water
www.biology.arizona.edu/biochemistry/tutorials/chemistry/page3.html
Chemistry of water and other aspects of biochemistry.

Chemistry: The Periodic Table Online
www.webelements.com
Detailed information about elements.

Chemistry Tutor
library.thinkquest.org/2923
A series of Web pages that help with chemistry assignments.

Chem4Kids
www.chem4Kids.com
Includes sections on matter, atoms, elements, and biochemistry.

Chemtutor Elements
www.chemtutor.com/elem.htm
Information on a selection of the elements.

Eric Weisstein's World of Chemistry
scienceworld.wolfram.com/chemistry
Chemistry information divided into eight broad topics, from chemical reactions to quantum chemistry.

General Chemistry Help
chemed.chem.purdue.edu/genchem
General information on chemistry plus movie clips of key concepts.

Molecular Models
chemlabs.uoregon.edu/GeneralResources/models/models.html
A site that explains the use of molecular models.

New Scientist
www.newscientist.com/home.ns
Online science magazine providing general news on scientific developments.

Periodic Tables
www.chemistrycoach.com/periodic_tables.htm#Periodic%20Tables
A list of links to sites that have information on the periodic table.

The Physical Properties of Minerals
mineral.galleries.com/minerals/physical.htm
Methods for identifying minerals.

Understanding Our Planet Through Chemistry
minerals.cr.usgs.gov/gips/aii-home.htm
Site that shows how chemists and geologists use analytical chemistry to study Earth.

Scientific American
www.sciam.com
Latest news on developments in science and technology.

Snowflakes and Snow Crystals
www.its.caltech.edu/~atomic/snowcrystals
A guide to snowflakes, snow crystals, and other ice phenomena.

Virtual Laboratory: Ideal Gas Laws
zebu.uoregon.edu/nsf/piston.html
University of Oregon site showing simulation of ideal gas laws.

What Is Salt?
www.saltinstitute.org/15.html
Information on common salt.

Periodic Table

The periodic table organizes all the chemical elements into a simple chart according to the physical and chemical properties of their atoms. The elements are arranged by atomic number from 1 to 116. The atomic number is based on the number of protons in the nucleus of the atom. The atomic mass is the combined mass of protons and neutrons in the nucleus. Each element has a chemical symbol that is an abbreviation of its name. In some cases, such as potassium,

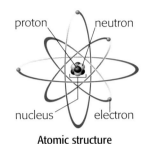

Atomic structure

33	—	Atomic (proton) number		
As	—	Chemical symbol		
Arsenic	—	Element name		
75	—	Atomic mass		

- ☐ HYDROGEN
- ☐ ALKALI METALS
- ☐ ALKALINE-EARTH METALS
- ☐ METALS
- ☐ LANTHANIDES

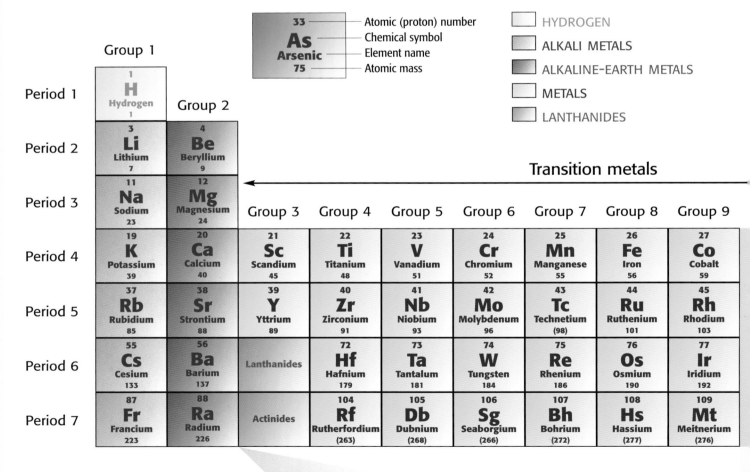

Transition metals

rare-earth elements ⎰ Lanthanides
         ⎱ Actinides

Group 1

Period 1 — 1 **H** Hydrogen 1

Group 2

Period 2 — 3 **Li** Lithium 7 | 4 **Be** Beryllium 9

Period 3 — 11 **Na** Sodium 23 | 12 **Mg** Magnesium 24

Group 3 · Group 4 · Group 5 · Group 6 · Group 7 · Group 8 · Group 9

Period 4 — 19 **K** Potassium 39 | 20 **Ca** Calcium 40 | 21 **Sc** Scandium 45 | 22 **Ti** Titanium 48 | 23 **V** Vanadium 51 | 24 **Cr** Chromium 52 | 25 **Mn** Manganese 55 | 26 **Fe** Iron 56 | 27 **Co** Cobalt 59

Period 5 — 37 **Rb** Rubidium 85 | 38 **Sr** Strontium 88 | 39 **Y** Yttrium 89 | 40 **Zr** Zirconium 91 | 41 **Nb** Niobium 93 | 42 **Mo** Molybdenum 96 | 43 **Tc** Technetium (98) | 44 **Ru** Ruthenium 101 | 45 **Rh** Rhodium 103

Period 6 — 55 **Cs** Cesium 133 | 56 **Ba** Barium 137 | Lanthanides | 72 **Hf** Hafnium 179 | 73 **Ta** Tantalum 181 | 74 **W** Tungsten 184 | 75 **Re** Rhenium 186 | 76 **Os** Osmium 190 | 77 **Ir** Iridium 192

Period 7 — 87 **Fr** Francium 223 | 88 **Ra** Radium 226 | Actinides | 104 **Rf** Rutherfordium (263) | 105 **Db** Dubnium (268) | 106 **Sg** Seaborgium (266) | 107 **Bh** Bohrium (272) | 108 **Hs** Hassium (277) | 109 **Mt** Meitnerium (276)

Lanthanides: 57 **La** Lanthanum 39 | 58 **Ce** Cerium 140 | 59 **Pr** Praseodymium 141 | 60 **Nd** Neodymium 144 | 61 **Pm** Promethium (145)

Actinides: 89 **Ac** Actinium 227 | 90 **Th** Thorium 232 | 91 **Pa** Protactinium 231 | 92 **U** Uranium 238 | 93 **Np** Neptunium (237)

the symbol is an abbreviation of its Latin name ("K" stands for *kalium*). The name by which the element is commonly known is given in full underneath the symbol. The last item in the element box is the atomic mass. This is the average mass of an atom of the element.

Scientists have arranged the elements into vertical columns called groups and horizontal rows called periods. Elements in any one group all have the same number of electrons in their outer shell and have similar chemical properties. Periods represent the increasing number of electrons it takes to fill the inner and outer shells and become stable. When all the spaces have been filled (Group 18 atoms have all their shells filled) the next period begins. Further explanation of the periodic table is given in Volume 5.

ACTINIDES

NOBLE GASES

NONMETALS

METALLOIDS

Group 18

			Group 13	Group 14	Group 15	Group 16	Group 17	2 **He** Helium 4
			5 **B** Boron 11	6 **C** Carbon 12	7 **N** Nitrogen 14	8 **O** Oxygen 16	9 **F** Fluorine 19	10 **Ne** Neon 20

Group 10 Group 11 Group 12

			13 **Al** Aluminum 27	14 **Si** Silicon 28	15 **P** Phosphorus 31	16 **S** Sulfur 32	17 **Cl** Chlorine 35	18 **Ar** Argon 40
28 **Ni** Nickel 59	29 **Cu** Copper 64	30 **Zn** Zinc 65	31 **Ga** Gallium 70	32 **Ge** Germanium 73	33 **As** Arsenic 75	34 **Se** Selenium 79	35 **Br** Bromine 80	36 **Kr** Krypton 84
46 **Pd** Palladium 106	47 **Ag** Silver 108	48 **Cd** Cadmium 112	49 **In** Indium 115	50 **Sn** Tin 119	51 **Sb** Antimony 122	52 **Te** Tellurium 128	53 **I** Iodine 127	54 **Xe** Xenon 131
78 **Pt** Platinum 195	79 **Au** Gold 197	80 **Hg** Mercury 201	81 **Tl** Thallium 204	82 **Pb** Lead 207	83 **Bi** Bismuth 209	84 **Po** Polonium (209)	85 **At** Astatine (210)	84 **Rn** Radon (222)
110 **Ds** Darmstadtium (281)	111 **Rg** Roentgenium (280)	112 **Uub** Ununbium (285)	113 **Uut** Ununtrium (284)	114 **Uuq** Ununquadium (289)	115 **Uup** Ununpentium (288)	116 **Uuh** Ununhexium (292)		

artificial elements

62 **Sm** Samarium 150	63 **Eu** Europium 152	64 **Gd** Gadolinium 157	65 **Tb** Terbium 159	66 **Dy** Dysprosium 163	67 **Ho** Holmium 165	68 **Er** Erbium 167	69 **Tm** Thulium 169	70 **Yb** Ytterbium 173	71 **Lu** Lutetium 175
94 **Pu** Plutonium (244)	95 **Am** Americium (243)	96 **Cm** Curium (247)	97 **Bk** Berkelium (247)	98 **Cf** Californium (251)	99 **Es** Einsteinium (252)	100 **Fm** Fermium (257)	101 **Md** Mendelevium (258)	102 **No** Nobelium (259)	103 **Lr** Lawrencium (260)

Glossary

acid Substance that dissolves in water to form hydrogen ions (H^+). Acids are neutralized by bases and alkalis and have a pH below 7.

alchemist Person who attempts to change one substance into another using a combination of primitive chemistry and magic.

alkali Substance that dissolves in water to form hydroxide ions (OH^-). Alkalis have a pH greater than 7 and will react with acids to form salts.

alkali metals Those metals that form Group 1 of the periodic table.

alkaline-earth metals Those metals that form Group 2 of the periodic table.

allotrope A different form of an element in which the atoms are arranged in a different structure.

alloy A metallic substance that contains two or more metals. An alloy may also be made of a metal and a small amount of a nonmetal. Steel, for example, is an alloy of iron and carbon.

alpha particle The nucleus of a helium atom. This particle has two protons and two neutrons.

amorphous Describes something that lacks a definite structure or shape.

atom The smallest independent building block of matter. All substances are made of atoms.

atomic mass number The number of protons and neutrons in an atom's nucleus.

atomic number The number of protons in a nucleus.

Avogadro's number The number of atoms, molecules, or ions in one mole of a pure substance. This number is 602,213,670,000,000,000,000, 000, or 6.0221367×10^{23}.

base Any substance that produces hydroxide ions (OH^-) is a base. All alkalis are bases.

beta particle An electron emitted by a decaying radioactive element.

boiling point The temperature at which a liquid turns into a gas.

bond A chemical connection between atoms.

buckminsterfullerene An allotrope of carbon shaped like a soccer ball.

by-product A substance that is produced when another material is made.

carbon dating Method for determining the age of a sample containing carbon.

chemical equation Symbols and numbers that show how reactants change into products during a chemical reaction.

chemical formula The letters and numbers that represent a chemical compound, such as "H_2O" for water.

chemical reaction The reaction of two or more chemicals (the reactants) to form new chemicals (the products).

chemical symbol The letters that represent a chemical, such as "Cl" for chlorine or "Na" for sodium.

combustion The reaction that causes burning. Combustion is generally a reaction with oxygen in the air.

compound Substance made from more than one element and that has undergone a chemical reaction.

compress To reduce in size or volume by squeezing or exerting pressure.

condensation The change of state from a gas to a liquid.

conductor A substance that carries electricity and heat.

corrosion The slow wearing away of metals or solids by chemical attack.

covalent bond Bond in which atoms share electrons.

crystal A solid made of regular repeating patterns of atoms.

crystal lattice The arrangement of atoms in a crystalline solid.

decay chain The sequence in which a radioactive element breaks down.

density The mass of substance in a unit of volume.

dipole attraction The attractive force between the electrically charged ends of molecules.

dissolve To form a solution.

distillation The process of evaporation and condensation used to separate a mixture of liquids according to their boiling points. Also a method of purifying a liquid.

electricity A stream of electrons or other charged particles moving through a substance.

electromagnetic radiation The energy emitted by a source, for example, X-rays, ultraviolet light, visible light, heat, or radio waves.

electromagnetic spectrum The range of energy waves that includes light, heat, and radio waves.

electron A tiny, negatively charged particle that moves around the nucleus of an atom.

electronegativity The power of an atom to attract an electron. Nonmetals, which have only a few spaces to fill in their outer shell, are the most electronegative elements. Metals, which have several empty spaces in their outer shells, tend to lose electrons in chemical reactions. Metals of this type are termed electropositive.

element A material that cannot be broken up into simpler ingredients. Elements contain only one type of atom.

endothermic reaction A reaction that absorbs heat energy.

energy The ability to do work.

energy level The shells around an atomic nucleus each represent a different energy level. Those closest to the nucleus have the lowest energy.

enthalpy The change in energy during a chemical reaction.

evaporation The change of state from a liquid to a gas when the liquid is at a temperature below its boiling point.

exothermic reaction A reaction that releases energy.

fission Process in which a large atom breaks up into two or more smaller fragments.

four elements The ancient theory that all matter consisted of only four elements (earth, air, fire, and water) and their combinations.

fusion Process by which two or more small atoms fuse to make a single larger atom.

gas State in which particles are not joined and are free to move in any direction.

heterogeneous mixture A mixture in which different substances are spread unevenly throughout.

homogeneous mixture A mixture in which one substance has dissolved or been completely mixed into another.

hydrogen bond A weak dipole attraction that always involves a hydrogen atom.

intermolecular bond The bonds that hold molecules together. These bonds are weaker than those between atoms in a molecule.

internal energy The total kinetic energy of all the particles in a system, plus all the chemical energy.

intramolecular bond Strong bond between atoms in a molecule.

ion An atom that has lost or gained one or more electrons and gained an electric charge.

ionic bond Bond in which one atom gives one or more electrons to another atom.

ionization The formation of ions by adding or removing electrons from atoms.

isotope Atoms of a given element must have the same number of protons but can have different numbers of neutrons. These different versions of the same element are called isotopes.

law of conservation of mass Law which states that in a chemical reaction, the total mass of the products is always equal to the total mass of the reactants.

liquid Substance in which particles are loosely bonded and are able to move freely around each other.

lubricant A substance that helps surfaces slide past each other.

malleable Describes a material that can be hammered into different shapes without breaking. Metals are malleable.

matter Anything that can be weighed.

melting point The temperature at which a solid changes into a liquid. When a liquid changes into a solid, this same temperature is also called the freezing point.

metal An element that is solid, shiny, malleable, ductile, and conductive.

metallic bond Bond in which outer electrons are free to move in the spaces between the atoms.

mixture Matter made from different types of substances that are not physically or chemically bonded together.

mole The amount of any substance that contains the same number of atoms as in 12 grams of carbon-12 atoms. This number is 6.022×10^{23}.

molecule Two or more bonded atoms that form a substance with specific properties.

neutron One of the particles that make up the nucleus of an atom. Neutrons do not have any electric charge.

noble gases A group of gases that rarely react with other elements.

nonmetal Any element that is not a metal. Most nonmetals are gases, such as hydrogen and argon.

nucleus The central part of an atom. The nucleus contains protons and neutrons. The exception is hydrogen, which contains only one proton.

organic A compound that contains chains or rings of carbon and hydrogen atoms.

particle The smallest portion or amount of something.

photon A particle that carries a quantity of energy, such as in the form of light.

plasma "Fourth state of matter" in which atoms have lost some or all of their electrons.

plastic An organic polymer that can be molded or shaped into objects or films by heat.

pressure The force produced by pressing on something.

product A new substance or substances created by a chemical reaction.

proton A positively charged particle found in an atom's nucleus.

radioactive decay The breakdown of an unstable nucleus through the loss of alpha and beta particles.

radiation The products of radioactivity—alpha and beta particles and gamma rays.

reactants The ingredients necessary for a chemical reaction.

relative atomic mass A measure of the mass of an atom compared with the mass of another atom. The values used are the same as those for atomic mass.

relative molecular mass The sum of all the atomic masses of the atoms in a molecule.

salt A compound made from positive and negative ions that forms when an alkali reacts with an acid.

shell The orbit of an electron. Each shell can contain a specific number of electrons and no more.

solid State of matter in which particles are held in a rigid arrangement.

state The form that matter takes—either a solid, a liquid, or a gas.

subatomic particles Particles that are smaller than an atom.

supernova The explosion of a very large star.

temperature A measure of how fast molecules are moving. It indicates the heat being given off by something.

valence electrons The electrons in the outer shell of an atom.

van der Waals forces Short-lived forces between atoms and molecules.

volume The space that a solid, liquid, or gas occupies.

Index